ARCHITECTS CONTRACTORS ENGINEERS

Guide To Construction COSTS

2009 Vol. XL

Architects Contractors Engineers Guide to Construction Costs 2009, Volume 40.

ISBN 978-1-58855-092-7

EDITOR'S NOTE 2009

This annually published book is designed to give a uniform estimating and cost control system to the General Building Contractor. It contains a complete system to be used with or without computers. It also contains Quick Estimating sections for preliminary conceptual budget estimates by Architects, Engineers and Contractors. Square Foot Estimating is also included for preliminary estimates.

The Metropolitan Area concept is also used and gives the cost modifiers to use for the variations between Metropolitan Areas. This encompasses over 50% of the industry. This book is published annually to be historically accurate with the traditional May-July wage contract settlements and to be a true construction year estimating and cost guide.

The Rate of Inflation in the Construction Industry in 2008 was *4%*. Labor contributed a *3%* increase and materials rose *5%*.

The Wage Rate for Skilled Trades increased an average of *3%* in 2008. Wage rates will probably increase at a *4%* average next year.

The Material Rate increased *4%* in 2008. The main increases were in asphalt, stone, iron pipe, steel and steel products, and copper products. Increases of double digits in some materials were offset by dramatic reductions again in the price of lumber and wood products.

Construction Volume should be slightly down or stay level 2009. Housing will decline yet again, and Industrial and Commercial Construction may rise. Highway and Heavy Construction should rise.

The Construction Industry has low to moderate inflation. Some materials should inflate at a slower pace, and some should be watched carefully in 2009.

We are recommending using an *4%* increase in your estimates for work beyond July 1, 2008.

Don Roth
Editor

CONTENTS

DIVISION 01 GENERAL..01
DIVISION 02 SITE WORK...02
DIVISION 03 CONCRETE..03
DIVISION 04 MASONRY...04
DIVISION 05 METALS..05
DIVISION 06 WOOD & PLASTICS...06
DIVISION 07 THERMAL & MOISTURE..07
DIVISION 08 DOORS AND WINDOWS...08
DIVISION 09 FINISHES..09
DIVISION 10 SPECIALTIES...10
DIVISION 11 EQUIPMENT...11
DIVISION 12 FURNISHINGS...12
DIVISION 13 SPECIAL CONSTRUCTION..13
DIVISION 14 CONVEYING SYSTEMS...14
DIVISION 15 MECHANICAL..15
DIVISION 16 ELECTRICAL..16
SQUARE FOOT COSTS...SF
AC&E ANNUAL COST INDEX..MM
INDEX...AI

Metropolitan Cost Index

The costs as presented in this book attempt to represent national averages. Costs, however, vary among regions, states and even between adjacent localities.

In order to more closely approximate the probable costs for specific locations throughout the U.S., this table of Metropolitan Cost Modifiers is provided in the following few pages. These adjustment factors are used to modify costs obtained from this book to help account for regional variations of construction costs and to provide a more accurate estimate for specific areas. The factors are formulated by comparing costs in a specific area to the costs as presented in the Costbook pages. An example of how to use these factors is shown below. Whenever local current costs are known, whether material prices or labor rates, they should be used when more accuracy is required.

Cost Obtained from Costbook Pages X **Metroploitan Cost Multiplier Divided by 100** = **Adjusted Cost**

For example, a project estimated to cost $125,000 using the Costbook pages can be adjusted to more closely approximate the cost in Los Angeles:

$$\$125,000 \quad X \quad \frac{115}{100} \quad = \quad \$143,750$$

State	Metropolitan Area	Multiplier		State	Metropolitan Area	Multiplier
AK	ANCHORAGE	137		FL	OCALA	95
AL	ANNISTON	90			ORLANDO	95
	AUBURN-OPELIKA	90			PANAMA CITY	83
	BIRMINGHAM	88			PENSACOLA	88
	DOTHAN	86			SARASOTA-BRADENTON	89
	GADSDEN	86			TALLAHASSEE	85
	HUNTSVILLE	88			TAMPA-ST. PETERSBURG-CLEARWATER	93
	MOBILE	90			WEST PALM BEACH-BOCA RATON	96
	MONTGOMERY	86		GA	ALBANY	85
	TUSCALOOSA	88			ATHENS	87
AR	FAYETTEVILLE-SPRINGDALE-ROGERS	80			ATLANTA	95
	FORT SMITH	86			AUGUSTA	83
	JONESBORO	85			COLUMBUS	84
	LITTLE ROCK-NORTH LITTLE ROCK	87			MACON	87
	PINE BLUFF	87			SAVANNAH	89
	TEXARKANA	86		HI	HONOLULU	135
AZ	FLAGSTAFF	98		IA	CEDAR RAPIDS	97
	PHOENIX-MESA	98			DAVENPORT	100
	TUCSON	97			DES MOINES	101
	YUMA	99			DUBUQUE	95
CA	BAKERSFIELD	112			IOWA CITY	100
	FRESNO	114			SIOUX CITY	95
	LOS ANGELES-LONG BEACH	115			WATERLOO-CEDAR FALLS	94
	MODESTO	111		ID	BOISE CITY	99
	OAKLAND	120			POCATELLO	97
	ORANGE COUNTY	114		IL	BLOOMINGTON-NORMAL	106
	REDDING	111			CHAMPAIGN-URBANA	105
	RIVERSIDE-SAN BERNARDINO	112			CHICAGO	115
	SACRAMENTO	114			DECATUR	104
	SALINAS	116			KANKAKEE	107
	SAN DIEGO	113			PEORIA-PEKIN	106
	SAN FRANCISCO	125			ROCKFORD	106
	SAN JOSE	122			SPRINGFIELD	104
	SAN LUIS OBISPO	110		IN	BLOOMINGTON	101
	SANTA CRUZ-WATSONVILLE	116			EVANSVILLE	100
	SANTA ROSA	117			FORT WAYNE	100
	STOCKTON-LODI	113			GARY	108
	VALLEJO-FAIRFIELD-NAPA	116			INDIANAPOLIS	104
	VENTURA	112			KOKOMO	100
	SANTA BARBARA	115			LAFAYETTE	100
CO	BOULDER-LONGMONT	96			MUNCIE	100
	COLORADO SPRINGS	101			SOUTH BEND	101
	DENVER	101			TERRE HAUTE	100
	FORT COLLINS-LOVELAND	94		KS	KANSAS CITY	99
	GRAND JUNCTION	96			LAWRENCE	94
	GREELEY	94			TOPEKA	93
	PUEBLO	98			WICHITA	93
CT	BRIDGEPORT	112		KY	LEXINGTON	93
	DANBURY	112			LOUISVILLE	96
	HARTFORD	111			OWENSBORO	94
	NEW HAVEN-MERIDEN	112		LA	ALEXANDRIA	88
	NEW LONDON-NORWICH	109			BATON ROUGE	92
	STAMFORD-NORWALK	115			HOUMA	92
	WATERBURY	111			LAFAYETTE	90
DC	WASHINGTON	104			LAKE CHARLES	92
DE	DOVER	104			MONROE	88
	WILMINGTON-NEWARK	105			NEW ORLEANS	95
FL	DAYTONA BEACH	92			SHREVEPORT-BOSSIER CITY	89
	FORT LAUDERDALE	96		MA	BARNSTABLE-YARMOUTH	115
	FORT MYERS-CAPE CORAL	88			BOSTON	118
	FORT PIERCE-PORT ST. LUCIE	95			BROCKTON	114
	FORT WALTON BEACH	97			FITCHBURG-LEOMINSTER	111
	GAINESVILLE	90			LAWRENCE	114
	JACKSONVILLE	94			LOWELL	109
	LAKELAND-WINTER HAVEN	90			NEW BEDFORD	114
	MELBOURNE-TITUSVILLE-PALM BAY	98			PITTSFIELD	110
	MIAMI	96			SPRINGFIELD	110
	NAPLES	98			WORCESTER	109

State	Metropolitan Area	Multiplier
MD	BALTIMORE	98
	CUMBERLAND	96
	HAGERSTOWN	91
ME	BANGOR	95
	LEWISTON-AUBURN	96
	PORTLAND	97
MI	ANN ARBOR	106
	DETROIT	111
	FLINT	105
	GRAND RAPIDS-MUSKEGON-HOLLAND	100
	JACKSON	104
	KALAMAZOO-BATTLE CREEK	96
	LANSING-EAST LANSING	104
	SAGINAW-BAY CITY-MIDLAND	102
MN	DULUTH	106
	MINNEAPOLIS-ST. PAUL	111
	ROCHESTER	106
	ST. CLOUD	108
MO	COLUMBIA	99
	JOPLIN	95
	KANSAS CITY	102
	SPRINGFIELD	97
	ST. JOSEPH	99
	ST. LOUIS	99
MS	BILOXI-GULFPORT-PASCAGOULA	87
	JACKSON	86
MT	BILLINGS	98
	GREAT FALLS	99
	MISSOULA	97
NC	ASHEVILLE	82
	CHARLOTTE	83
	FAYETTEVILLE	84
	GREENSBORO-WINSTON-SALEM-H POINT	83
	GREENVILLE	83
	HICKORY-MORGANTON-LENOIR	79
	RALEIGH-DURHAM-CHAPEL HILL	83
	ROCKY MOUNT	79
	WILMINGTON	84
ND	BISMARCK	95
	FARGO	96
	GRAND FORKS	94
NE	LINCOLN	92
	OMAHA	96
NH	MANCHESTER	100
	NASHUA	98
	PORTSMOUTH	92
NJ	ATLANTIC-CAPE MAY	110
	BERGEN-PASSAIC	113
	JERSEY CITY	112
	MIDDLESEX-SOMERSET-HUNTERDON	108
	MONMOUTH-OCEAN	111
	NEWARK	114
	TRENTON	111
	VINELAND-MILLVILLE-BRIDGETON	109
NM	ALBUQUERQUE	96
	LAS CRUCES	92
	SANTA FE	98
NV	LAS VEGAS	108
	RENO	107
NY	ALBANY-SCHENECTADY-TROY	102
	BINGHAMTON	100
	BUFFALO-NIAGARA FALLS	108
	ELMIRA	95
	GLENS FALLS	94
	JAMESTOWN	102
	NASSAU-SUFFOLK	115
	NEW YORK	132
	ROCHESTER	106
	SYRACUSE	103
	UTICA-ROME	103

State	Metropolitan Area	Multiplier
OH	AKRON	103
	CANTON-MASSILLON	101
	CINCINNATI	98
	CLEVELAND-LORAIN-ELYRIA	106
	COLUMBUS	101
	DAYTON-SPRINGFIELD	99
	LIMA	101
	MANSFIELD	97
	STEUBENVILLE	104
	TOLEDO	103
	YOUNGSTOWN-WARREN	100
OK	ENID	88
	LAWTON	89
	OKLAHOMA CITY	91
	TULSA	90
OR	EUGENE-SPRINGFIELD	106
	MEDFORD-ASHLAND	104
	PORTLAND	108
	SALEM	107
PA	ALLENTOWN-BETHLEHEM-EASTON	105
	ALTOONA	103
	ERIE	103
	HARRISBURG-LEBANON-CARLISLE	101
	JOHNSTOWN	104
	LANCASTER	101
	PHILADELPHIA	114
	PITTSBURGH	104
	READING	103
	SCRANTON-WILKES-BARRE-HAZLETON	102
	STATE COLLEGE	98
	WILLIAMSPORT	100
	YORK	102
RI	PROVIDENCE	109
SC	AIKEN	91
	CHARLESTON-NORTH CHARLESTON	85
	COLUMBIA	85
	FLORENCE	82
	GREENVILLE-SPARTANBURG-ANDERSON	84
	MYRTLE BEACH	90
SD	RAPID CITY	89
	SIOUX FALLS	90
TN	CHATTANOOGA	88
	JACKSON	87
	JOHNSON CITY	84
	KNOXVILLE	88
	MEMPHIS	93
	NASHVILLE	92
TX	ABILENE	86
	AMARILLO	87
	AUSTIN-SAN MARCOS	88
	BEAUMONT-PORT ARTHUR	88
	BROWNSVILLE-HARLINGEN-SAN BENITO	89
	BRYAN-COLLEGE STATION	87
	CORPUS CHRISTI	85
	DALLAS	91
	EL PASO	85
	FORT WORTH-ARLINGTON	91
	GALVESTON-TEXAS CITY	90
	HOUSTON	92
	LAREDO	78
	LONGVIEW-MARSHALL	83
	LUBBOCK	88
	MCALLEN-EDINBURG-MISSION	84
	ODESSA-MIDLAND	85
	SAN ANGELO	83
	SAN ANTONIO	89
	TEXARKANA	86
	TYLER	86
	VICTORIA	86
	WACO	86

State	Metropolitan Area	Multiplier	State	Metropolitan Area	Multiplier
UT	PROVO-OREM	94	WI	APPLETON-OSHKOSH-NEENAH	101
	SALT LAKE CITY-OGDEN	93		EAU CLAIRE	100
VA	CHARLOTTESVILLE	88		GREEN BAY	101
	LYNCHBURG	87		JANESVILLE-BELOIT	101
	NORFOLK-VA BEACH-NEWPORT NEWS	91		KENOSHA	103
	RICHMOND-PETERSBURG	91		LA CROSSE	99
	ROANOKE	85		MADISON	100
VT	BURLINGTON	98		MILWAUKEE-WAUKESHA	107
WA	BELLINGHAM	112		RACINE	103
	BREMERTON	110		WAUSAU	100
	OLYMPIA	108	WV	CHARLESTON	99
	RICHLAND-KENNEWICK-PASCO	106		HUNTINGTON	100
	SEATTLE-BELLEVUE-EVERETT	111		PARKERSBURG	98
	SPOKANE	108		WHEELING	100
	TACOMA	111	WY	CASPER	92
	YAKIMA	106		CHEYENNE	93

HOW TO USE THIS BOOK

Labor Unit Columns
- ➤ Units *do not include* Workers Comp., Unemployment, and FICA on labor (approx. 35%).
- ➤ Units *do not include* general conditions and equipment (approx. 10%).
- ➤ Units *do not include* contractors' overhead and profit (approx. 10%).
- ➤ Units are Union or Government Minimum Area wages.

Material Unit Columns
- ➤ Units *do not include* general conditions and equipment (approx. 10%).
- ➤ Units *do not include* sales or use taxes (approx. 5%) of material cost.
- ➤ Units *do not include* contractors' overhead and profit (approx. 10%) and are FOB job site.

Cost Unit Columns (Subcontractors)
- ➤ Units *do not include* general contractors' overhead or profit.

Quick Estimating Sections - For Preliminary and Conceptual Estimating
- ➤ Includes all labor, material, general conditions, equipment, taxes.

Construction Modifiers - For Metro Area Cost Variation Adjustments
- ➤ Labor items are percentage increases or decreases from 100% for labor units used in the book's labor unit columns.
- ➤ Material and Labor combined are percentage increases and decreases from 100% for the book's Unit Cost columns, Quick Estimating sections and Sq.Ft. Cost Section.

EXAMPLE:		UNIT	LABOR	MATERIAL	COST
2" x 4" Wood Studs & Plates - 8' @ 16" O.C.		BdFt	.74	.45	1.19
Add 35% Taxes & Insurance on Labor			.26	-	.26
Add 5% Material Sales Tax			-	.03	.03
					1.48
Add 10% General Conditions & Equipment					.15
					1.63
Add 10% Overhead & Profit					.17
BdFt Cost - Bid Type Estimating				BdFt	1.80
or SqFt Wall Cost with 3 Plates - Quick Estimating (page 6A-14)				or SqFt	1.36

Purchasing Practice
- L&M - Labor and Material Subcontractor - Firm bid to G.C. according to plans and specifications for all labor and material in that section or subsection.
- M - Material Subcontractor - Bid to G.C.
- L - Labor Subcontractor - Bid to G.C.

TABLE OF CONTENTS   PAGE

01020.10 - ALLOWANCES	01-2
01050.10 - FIELD STAFF	01-2
01310.10 - SCHEDULING	01-3
01410.10 - TESTING	01-3
01500.10 - TEMPORARY FACILITIES	01-3
01505.10 - MOBILIZATION	01-3
01525.10 - CONSTRUCTION AIDS	01-4
01525.20 - TEMPORARY CONST. SHELTERS	01-4
01570.10 - SIGNS	01-4
01600.10 - EQUIPMENT	01-4
01740.10 - BONDS	01-6

		UNIT	LABOR	MAT.	TOTAL
01020.10	**ALLOWANCES**				
0090	Overhead				
1000	$20,000 project				
1040	Average	PCT.			20.00
1080	$100,000 project				
1120	Average	PCT.			15.00
1160	$500,000 project				
1180	Average	PCT.			12.00
1220	$1,000,000 project				
1260	Average	PCT.			10.00
1480	Profit				
1500	$20,000 project				
1540	Average	PCT.			15.00
1580	$100,000 project				
1620	Average	PCT.			12.00
1660	$500,000 project				
1700	Average	PCT.			10.00
1740	$1,000,000 project				
1780	Average	PCT.			8.00
2000	Professional fees				
2100	Architectural				
2120	$100,000 project				
2160	Average	PCT.			10.00
2200	$500,000 project				
2240	Average	PCT.			8.00
2280	$1,000,000 project				
2320	Average	PCT.			7.00
4080	Taxes				
5000	Sales tax				
5040	Average	PCT.			5.00
5080	Unemployment				
5120	Average	PCT.			6.50
5200	Social security (FICA)	"			7.85
01050.10	**FIELD STAFF**				
1000	Superintendent				
1020	Minimum	YEAR			75,600
1040	Average	"			110,250
1060	Maximum	"			157,500
1080	Field engineer				
1100	Minimum	YEAR			57,750
1120	Average	"			89,250
1140	Maximum	"			131,250
1160	Foreman				
1180	Minimum	YEAR			42,000
1200	Average	"			68,250
1220	Maximum	"			99,750
1240	Bookkeeper/timekeeper				
1260	Minimum	YEAR			24,150
1280	Average	"			31,500
1300	Maximum	"			52,500
1320	Watchman				
1340	Minimum	YEAR			16,800

		UNIT	LABOR	MAT.	TOTAL
01050.10	**FIELD STAFF, Cont'd...**				
1360	Average	YEAR			21,000
1380	Maximum	"			33,600
01310.10	**SCHEDULING**				
0090	Scheduling for				
1000	$100,000 project				
1040	Average	PCT.			2.10
1080	$500,000 project				
1120	Average	PCT.			1.05
1160	$1,000,000 project				
1200	Average	PCT.			0.84
4000	Scheduling software				
4020	Minimum	EA.			400
4040	Average	"			2,310
4060	Maximum	"			46,200
01410.10	**TESTING**				
1080	Testing concrete, per test				
1100	Minimum	EA.			19.00
1120	Average	"			31.75
1140	Maximum	"			64.00
1160	Soil, per test				
1180	Minimum	EA.			38.25
1200	Average	"			96.00
1220	Maximum	"			250
01500.10	**TEMPORARY FACILITIES**				
1000	Barricades, temporary				
1010	Highway				
1020	Concrete	L.F.	3.75	11.25	15.00
1040	Wood	"	1.50	3.67	5.17
1060	Steel	"	1.25	3.93	5.18
1080	Pedestrian barricades				
1100	Plywood	S.F.	1.25	2.62	3.87
1120	Chain link fence	"	1.25	2.95	4.20
2000	Trailers, general office type, per month				
2020	Minimum	EA.			200
2040	Average	"			330
2060	Maximum	"			660
2080	Crew change trailers, per month				
2100	Minimum	EA.			120
2120	Average	"			130
2140	Maximum	"			200
01505.10	**MOBILIZATION**				
1000	Equipment mobilization				
1020	Bulldozer				
1040	Minimum	EA.			180
1060	Average	"			380
1080	Maximum	"			630
1100	Backhoe/front-end loader				
1120	Minimum	EA.			110
1140	Average	"			190
1160	Maximum	"			410
1180	Crane, crawler type				

		UNIT	LABOR	MAT.	TOTAL
01505.10	**MOBILIZATION, Cont'd...**				
1200	Minimum	EA.			1,990
1220	Average	"			4,880
1240	Maximum	"			10,480
1260	Truck crane				
1280	Minimum	EA.			470
1300	Average	"			700
1320	Maximum	"			1,210
1340	Pile driving rig				
1360	Minimum	EA.			9,030
1380	Average	"			18,070
1400	Maximum	"			32,530
01525.10	**CONSTRUCTION AIDS**				
1000	Scaffolding/staging, rent per month				
1020	Measured by lineal feet of base				
1040	10' high	L.F.			11.50
1060	20' high	"			20.75
1080	30' high	"			29.00
1100	40' high	"			33.50
1120	50' high	"			40.00
1140	Measured by square foot of surface				
1160	Minimum	S.F.			0.51
1180	Average	"			0.87
1200	Maximum	"			1.57
1220	Safety nets, heavy duty, per job				
1240	Minimum	S.F.			0.33
1260	Average	"			0.40
1280	Maximum	"			0.89
1300	Tarpaulins, fabric, per job				
1320	Minimum	S.F.			0.23
1340	Average	"			0.40
1360	Maximum	"			1.03
01525.20	**TEMPORARY CONST. SHELTERS**				
0010	Standard, alum. with fabric, 12'x20'x15' Ht.	S.F.			15.50
0020	12'x20'x20' Ht.	"			18.00
01570.10	**SIGNS**				
0080	Construction signs, temporary				
1000	Signs, 2' x 4'				
1020	Minimum	EA.			33.25
1040	Average	"			80.00
1060	Maximum	"			280
1160	Signs, 8' x 8'				
1180	Minimum	EA.			90.00
1200	Average	"			280
1220	Maximum	"			2,900
01600.10	**EQUIPMENT**				
0080	Air compressor				
1000	60 cfm				
1020	By day	EA.			88.00
1030	By week	"			260
1040	By month	"			790
1200	600 cfm				

		UNIT	LABOR	MAT.	TOTAL
01600.10	**EQUIPMENT, Cont'd...**				
1210	By day	EA.			510
1220	By week	"			1,520
1230	By month	"			4,560
1300	Air tools, per compressor, per day				
1310	Minimum	EA.			33.75
1320	Average	"			42.25
1330	Maximum	"			59.00
1400	Generators, 5 kw				
1410	By day	EA.			85.00
1420	By week	"			250
1430	By month	"			770
1500	Heaters, salamander type, per week				
1510	Minimum	EA.			100
1520	Average	"			140
1530	Maximum	"			300
1600	Pumps, submersible				
1605	50 gpm				
1610	By day	EA.			68.00
1620	By week	"			200
1630	By month	"			610
1675	500 gpm				
1680	By day	EA.			140
1690	By week	"			400
1700	By month	"			1,220
1900	Diaphragm pump, by week				
1920	Minimum	EA.			120
1930	Average	"			200
1940	Maximum	"			420
2000	Pickup truck				
2020	By day	EA.			130
2030	By week	"			370
2040	By month	"			1,150
2080	Dump truck				
2100	6 cy truck				
2120	By day	EA.			340
2130	By week	"			1,010
2140	By month	"			3,040
2300	16 cy truck				
2310	By day	EA.			670
2320	By week	"			2,030
2340	By month	"			6,080
2400	Backhoe, track mounted				
2420	1/2 cy capacity				
2430	By day	EA.			690
2440	By week	"			2,110
2450	By month	"			6,250
2500	1 cy capacity				
2510	By day	EA.			1,100
2520	By week	"			3,290
2530	By month	"			9,880
2600	3 cy capacity				
2620	By day	EA.			3,550

		UNIT	LABOR	MAT.	TOTAL
01600.10	**EQUIPMENT, Cont'd...**				
2640	By week	EA.			10,640
2680	By month	"			31,940
3000	Backhoe/loader, rubber tired				
3005	1/2 cy capacity				
3010	By day	EA.			420
3020	By week	"			1,270
3030	By month	"			3,800
3035	3/4 cy capacity				
3040	By day	EA.			510
3050	By week	"			1,520
3060	By month	"			4,560
3200	Bulldozer				
3205	75 hp				
3210	By day	EA.			590
3220	By week	"			1,770
3230	By month	"			5,320
4000	Cranes, crawler type				
4005	15 ton capacity				
4010	By day	EA.			760
4020	By week	"			2,280
4030	By month	"			6,840
4070	50 ton capacity				
4080	By day	EA.			1,690
4090	By week	"			5,070
4100	By month	"			15,210
4145	Truck mounted, hydraulic				
4150	15 ton capacity				
4160	By day	EA.			720
4170	By week	"			2,150
4180	By month	"			6,210
5380	Loader, rubber tired				
5385	1 cy capacity				
5390	By day	EA.			510
5400	By week	"			1,520
5410	By month	"			4,560
7000	Scraper				
7010	Elevated scraper, not including bulldozer, 12 c.y.				
7020	By day	EA.			1,160
7030	By week	"			3,460
7040	By month	"			9,820
7100	Self-propelled scraper, 14 c.y.				
7110	By day	EA.			2,310
7120	By week	"			6,930
7130	By month	"			19,630
01740.10	**DOND3**				
1000	Performance bonds				
1020	Minimum	PCT.			0.64
1040	Average	"			2.01
1060	Maximum	"			3.18

TABLE OF CONTENTS   PAGE

02010.10 - SOIL BORING	02-2
02020.10 - BLASTING ROCK	02-2
02060.10 - BUILDING DEMOLITION	02-2
02065.15 - SAW CUTTING PAVEMENT	02-4
02075.90 - TORCH CUTTING	02-4
02105.20 - FENCES	02-4
02110.50 - TREE CUTTING & CLEARING	02-5
02210.10 - HAULING MATERIAL	02-5
02210.30 - BULK EXCAVATION	02-5
02220.10 - BORROW	02-6
02220.40 - BUILDING EXCAVATION	02-7
02220.50 - UTILITY EXCAVATION	02-7
02220.60 - TRENCHING	02-7
02220.70 - ROADWAY EXCAVATION	02-8
02220.71 - BASE COURSE	02-8
02220.90 - HAND EXCAVATION	02-8
02240.05 - SOIL STABILIZATION	02-9
02270.40 - RIPRAP	02-9
02280.20 - SOIL TREATMENT	02-9
02290.30 - WEED CONTROL	02-9
02360.50 - PRESTRESSED PILING	02-9
02360.60 - STEEL PILES	02-9
02360.65 - STEEL PIPE PILES	02-9
02360.80 - WOOD AND TIMBER PILES	02-10
02510.20 - ASPHALT SURFACES	02-10
02520.10 - CONCRETE PAVING	02-10
02605.30 - MANHOLES	02-10
02610.40 - DUCTILE IRON PIPE	02-11
02610.60 - PLASTIC PIPE	02-11
02610.90 - VITRIFIED CLAY PIPE	02-12
02670.10 - WELLS	02-12
02690.10 - STORAGE TANKS	02-12
02720.70 - UNDERDRAIN	02-12
02730.10 - SANITARY SEWERS	02-13
02740.10 - DRAINAGE FIELDS	02-13
02740.50 - SEPTIC TANKS	02-13
02810.40 - LAWN IRRIGATION	02-13
02830.10 - CHAIN LINK FENCE	02-13
02830.20 - WOOD FENCE	02-13
02830.70 - RECREATIONAL COURTS	02-14
02840.40 - PARKING BARRIERS	02-14
02910.10 - SHRUB & TREE MAINTENANCE	02-14
02920.10 - TOPSOIL	02-14
02930.30 - SEEDING	02-15
02970.10 - FERTILIZING	02-15
02980.10 - LANDSCAPE ACCESSORIES	02-15
QUICK ESTIMATING	**02A-17**

		UNIT	LABOR	MAT.	TOTAL
02010.10	**SOIL BORING**				
1000	Borings, uncased, stable earth				
1020	2-1/2" dia.	L.F.	26.50		26.50
1040	4" dia.	"	30.25		30.25
1500	Cased, including samples				
1520	2-1/2" dia.	L.F.	35.25		35.25
1540	4" dia.	"	60.00		60.00
02020.10	**BLASTING ROCK**				
0010	Blasting mats, up to 1500 c.y.	C.Y.	87.00	2.75	89.75
0020	Buried explosives, up to 1500 c.y.	"	26.25	2.75	29.00
02060.10	**BUILDING DEMOLITION**				
0090	Building, complete with disposal				
0200	Wood frame	C.F.	0.30		0.30
0300	Concrete	"	0.46		0.46
0400	Steel frame	"	0.61		0.61
1000	Partition removal				
1100	Concrete block partitions				
1120	4" thick	S.F.	1.87		1.87
1140	8" thick	"	2.50		2.50
1160	12" thick	"	3.41		3.41
1200	Brick masonry partitions				
1220	4" thick	S.F.	1.87		1.87
1240	8" thick	"	2.34		2.34
1260	12" thick	"	3.12		3.12
1280	16" thick	"	4.69		4.69
1380	Cast in place concrete partitions				
1400	Unreinforced				
1421	6" thick	S.F.	14.00		14.00
1423	8" thick	"	15.00		15.00
1425	10" thick	"	17.50		17.50
1427	12" thick	"	21.25		21.25
1440	Reinforced				
1441	6" thick	S.F.	16.25		16.25
1443	8" thick	"	21.25		21.25
1445	10" thick	"	23.50		23.50
1447	12" thick	"	28.25		28.25
1500	Terra cotta				
1520	To 6" thick	S.F.	1.87		1.87
1700	Stud partitions				
1720	Metal or wood, with drywall both sides	S.F.	1.87		1.87
1740	Metal studs, both sides, lath and plaster	"	2.50		2.50
2000	Door and frame removal				
2020	Hollow metal in masonry wall				
2030	Single				
2040	2'6"x6'8"	EA.	47.00		47.00
2060	3'x7'	"	63.00		63.00
2070	Double				
2080	3'x7'	EA.	75.00		75.00
2085	4'x8'	"	75.00		75.00
2140	Wood in framed wall				
2150	Single				
2160	2'6"x6'8"	EA.	26.75		26.75

02060.10 BUILDING DEMOLITION, Cont'd...

		UNIT	LABOR	MAT.	TOTAL
2180	3'x6'8"	EA.	31.25		31.25
2190	Double				
2200	2'6"x6'8"	EA.	37.50		37.50
2220	3'x6'8"	"	41.75		41.75
2240	Remove for re-use				
2260	Hollow metal	EA.	94.00		94.00
2280	Wood	"	63.00		63.00
2300	Floor removal				
2340	Brick flooring	S.F.	1.50		1.50
2360	Ceramic or quarry tile	"	0.83		0.83
2380	Terrazzo	"	1.66		1.66
2400	Heavy wood	"	1.00		1.00
2420	Residential wood	"	1.07		1.07
2440	Resilient tile or linoleum	"	0.37		0.37
2500	Ceiling removal				
2520	Acoustical tile ceiling				
2540	Adhesive fastened	S.F.	0.37		0.37
2560	Furred and glued	"	0.31		0.31
2580	Suspended grid	"	0.23		0.23
2600	Drywall ceiling				
2620	Furred and nailed	S.F.	0.41		0.41
2640	Nailed to framing	"	0.37		0.37
2660	Plastered ceiling				
2680	Furred on framing	S.F.	0.93		0.93
2700	Suspended system	"	1.25		1.25
2800	Roofing removal				
2820	Steel frame				
2840	Corrugated metal roofing	S.F.	0.75		0.75
2860	Built-up roof on metal deck	"	1.25		1.25
2900	Wood frame				
2920	Built up roof on wood deck	S.F.	1.15		1.15
2940	Roof shingles	"	0.62		0.62
2960	Roof tiles	"	1.25		1.25
8900	Concrete frame	C.F.	2.50		2.50
8920	Concrete plank	S.F.	1.87		1.87
8940	Built-up roof on concrete	"	1.07		1.07
9200	Cut-outs				
9230	Concrete, elevated slabs, mesh reinforcing				
9240	Under 5 cf	C.F.	37.50		37.50
9260	Over 5 cf	"	31.25		31.25
9270	Bar reinforcing				
9280	Under 5 cf	C.F.	63.00		63.00
9290	Over 5 cf	"	47.00		47.00
9300	Window removal				
9301	Metal windows, trim included				
9302	2'x3'	EA.	37.50		37.50
9304	2'x4'	"	41.75		41.75
9306	2'x6'	"	47.00		47.00
9308	3'x4'	"	47.00		47.00
9310	3'x6'	"	54.00		54.00
9312	3'x8'	"	63.00		63.00
9314	4'x4'	"	63.00		63.00

		UNIT	LABOR	MAT.	TOTAL
02060.10	**BUILDING DEMOLITION, Cont'd...**				
9315	4'x6'	EA.	75.00		75.00
9316	4'x8'	"	94.00		94.00
9317	Wood windows, trim included				
9318	2'x3'	EA.	20.75		20.75
9319	2'x4'	"	22.00		22.00
9320	2'x6'	"	23.50		23.50
9321	3'x4'	"	25.00		25.00
9322	3'x6'	"	26.75		26.75
9324	3'x8'	"	28.75		28.75
9325	6'x4'	"	31.25		31.25
9326	6'x6'	"	34.00		34.00
9327	6'x8'	"	37.50		37.50
9329	Walls, concrete, bar reinforcing				
9330	Small jobs	C.F.	25.00		25.00
9340	Large jobs	"	20.75		20.75
9360	Brick walls, not including toothing				
9390	4" thick	S.F.	1.87		1.87
9400	8" thick	"	2.34		2.34
9410	12" thick	"	3.12		3.12
9415	16" thick	"	4.69		4.69
9420	Concrete block walls, not including toothing				
9440	4" thick	S.F.	2.08		2.08
9450	6" thick	"	2.20		2.20
9460	8" thick	"	2.34		2.34
9465	10" thick	"	2.68		2.68
9470	12" thick	"	3.12		3.12
9500	Rubbish handling				
9519	Load in dumpster or truck				
9520	Minimum	C.F.	0.83		0.83
9540	Maximum	"	1.25		1.25
9550	For use of elevators, add				
9560	Minimum	C.F.	0.18		0.18
9570	Maximum	"	0.37		0.37
9600	Rubbish hauling				
9640	Hand loaded on trucks, 2 mile trip	C.Y.	32.75		32.75
9660	Machine loaded on trucks, 2 mile trip	"	21.25		21.25
02065.15	**SAW CUTTING PAVEMENT**				
0100	Pavement, bituminous				
0110	2" thick	L.F.	1.64		1.64
0120	3" thick	"	2.05		2.05
0200	Concrete pavement, with wire mesh				
0210	4" thick	L.F.	3.15		3.15
0212	5" thick	"	3.42		3.42
0300	Plain concrete, unreinforced				
0320	4" thick	L.F.	2.73		2.73
0340	5" thick	"	3.15		3.15
02075.90	**TORCH CUTTING**				
0010	Steel plate, 1/2" thick,	L.F.	0.82	0.49	1.31
0020	1" thick	"	1.65	0.71	2.36

DIVISION # 02 SITEWORK

		UNIT	LABOR	MAT.	TOTAL
02105.20	**FENCES**				
0060	Remove fencing				
0080	Chain link, 8' high				
0100	For disposal	L.F.	1.87		1.87
0200	For reuse	"	4.69		4.69
0980	Wood				
1000	4' high	S.F.	1.25		1.25
1960	Masonry				
1980	8" thick				
2000	4' high	S.F.	3.75		3.75
2020	6' high	"	4.69		4.69
02110.50	**TREE CUTTING & CLEARING**				
0980	Cut trees and clear out stumps				
1000	9" to 12" dia.	EA.	420		420
1400	To 24" dia.	"	530		530
1600	24" dia. and up	"	700		700
02210.10	**HAULING MATERIAL**				
0090	Haul material by 10 cy dump truck, round trip distance				
0100	1 mile	C.Y.	4.56		4.56
0110	2 mile	"	5.47		5.47
0120	5 mile	"	7.46		7.46
0130	10 mile	"	8.21		8.21
0140	20 mile	"	9.12		9.12
0150	30 mile	"	11.00		11.00
2000	Site grading, cut & fill, sandy clay, 200' haul, 75 hp dozer	"	3.28		3.28
6000	Spread topsoil by equipment on site	"	3.64		3.64
6980	Site grading (cut and fill to 6") less than 1 acre				
7000	75 hp dozer	C.Y.	5.47		5.47
7600	1.5 cy backhoe/loader	"	8.21		8.21
02210.30	**BULK EXCAVATION**				
1000	Excavation, by small dozer				
1020	Large areas	C.Y.	1.64		1.64
1040	Small areas	"	2.73		2.73
1060	Trim banks	"	4.10		4.10
1700	Hydraulic excavator				
1720	1 cy capacity				
1740	Light material	C.Y.	3.52		3.52
1760	Medium material	"	4.22		4.22
1780	Wet material	"	5.28		5.28
1790	Blasted rock	"	6.03		6.03
1800	1-1/2 cy capacity				
1820	Light material	C.Y.	1.41		1.41
1840	Medium material	"	1.88		1.88
1860	Wet material	"	2.26		2.26
2000	Wheel mounted front-end loader				
2020	7/8 cy capacity				
2040	Light material	C.Y.	2.83		2.83
2060	Medium material	"	3.23		3.23
2080	Wet material	"	3.77		3.77
2100	Blasted rock	"	4.53		4.53
2300	2-1/2 cy capacity				
2320	Light material	C.Y.	1.33		1.33

© 2008 By Design & Construction Resources

02 - 5

		UNIT	LABOR	MAT.	TOTAL
02210.30	**BULK EXCAVATION, Cont'd...**				
2340	Medium material	C.Y.	1.41		1.41
2360	Wet material	"	1.51		1.51
2380	Blasted rock	"	1.61		1.61
2600	Track mounted front-end loader				
2620	1-1/2 cy capacity				
2640	Light material	C.Y.	1.88		1.88
2660	Medium material	"	2.05		2.05
2680	Wet material	"	2.26		2.26
2700	Blasted rock	"	2.51		2.51
2720	2-3/4 cy capacity				
2740	Light material	C.Y.	1.13		1.13
2760	Medium material	"	1.25		1.25
2780	Wet material	"	1.41		1.41
2790	Blasted rock	"	1.61		1.61
4000	Scraper -500' haul				
4010	Elevated scraper, not including bulldozer, 12 c.y.				
4020	Light material	C.Y.	3.77		3.77
4030	Medium material	"	4.11		4.11
4040	Wet material	"	4.53		4.53
4050	Blasted rock	"	5.03		5.03
4100	Self-propelled scraper, 14 c.y.				
4110	Light material	C.Y.	3.48		3.48
4120	Medium material	"	3.77		3.77
4130	Wet material	"	4.11		4.11
4140	Blasted rock	"	4.53		4.53
4200	1,000' haul				
4210	Elevated scraper, not including bulldozer, 12 c.y.				
4220	Light material	C.Y.	4.53		4.53
4230	Medium material	"	5.03		5.03
4240	Wet material	"	5.66		5.66
4250	Blasted rock	"	6.47		6.47
4300	Self-propelled scraper, 14 c.y.				
4310	Light material	C.Y.	4.11		4.11
4320	Medium material	"	4.53		4.53
4330	Wet material	"	5.03		5.03
4340	Blasted rock	"	5.66		5.66
4400	2,000' haul				
4410	Elevated scraper, not including bulldozer, 12 c.y.				
4420	Light material	C.Y.	5.66		5.66
4430	Medium material	"	6.47		6.47
4440	Wet material	"	7.55		7.55
4450	Blasted rock	"	9.06		9.06
4510	Self-propelled scraper, 14 c.y.				
4520	Light material	C.Y.	5.03		5.03
4530	Medium material	"	5.66		5.66
4540	Wet material	"	6.47		6.47
4550	Blasted rock	"	7.55		7.55
02220.10	**BORROW**				
1000	Borrow fill, F.O.B. at pit				
1005	Sand, haul to site, round trip				
1010	10 mile	C.Y.	11.25	15.50	26.75

		UNIT	LABOR	MAT.	TOTAL
02220.10	**BORROW, Cont'd...**				
1020	20 mile	C.Y.	18.75	15.50	34.25
1030	30 mile	"	28.25	15.50	43.75
3980	Place borrow fill and compact				
4000	Less than 1 in 4 slope	C.Y.	5.66	15.50	21.16
4100	Greater than 1 in 4 slope	"	7.55	15.50	23.05
02220.40	**BUILDING EXCAVATION**				
0090	Structural excavation, unclassified earth				
0100	3/8 cy backhoe	C.Y.	15.00		15.00
0110	3/4 cy backhoe	"	11.25		11.25
0120	1 cy backhoe	"	9.44		9.44
0600	Foundation backfill and compaction by machine	"	22.75		22.75
02220.50	**UTILITY EXCAVATION**				
2080	Trencher, sandy clay, 8" wide trench				
2100	18" deep	L.F.	1.82		1.82
2200	24" deep	"	2.05		2.05
2300	36" deep	"	2.34		2.34
6080	Trench backfill, 95% compaction				
7000	Tamp by hand	C.Y.	23.50		23.50
7050	Vibratory compaction	"	18.75		18.75
7060	Trench backfilling, with borrow sand, place & compact	"	18.75	12.25	31.00
02220.60	**TRENCHING**				
0100	Trenching and continuous footing excavation				
0980	By gradall				
1000	1 cy capacity				
1040	Medium soil	C.Y.	3.48		3.48
1080	Loose rock	"	4.11		4.11
1090	Blasted rock	"	4.35		4.35
1095	By hydraulic excavator				
1100	1/2 cy capacity				
1140	Medium soil	C.Y.	4.11		4.11
1180	Loose rock	"	5.03		5.03
1190	Blasted rock	"	5.66		5.66
1200	1 cy capacity				
1240	Medium soil	C.Y.	2.83		2.83
1280	Loose rock	"	3.23		3.23
1300	Blasted rock	"	3.48		3.48
1600	2 cy capacity				
1640	Medium soil	C.Y.	2.38		2.38
1680	Loose rock	"	2.66		2.66
1690	Blasted rock	"	2.83		2.83
3000	Hand excavation				
3100	Bulk, wheeled 100'				
3120	Normal soil	C.Y.	41.75		41.75
3140	Sand or gravel	"	37.50		37.50
3160	Medium clay	"	54.00		54.00
3180	Heavy clay	"	75.00		75.00
3200	Loose rock	"	94.00		94.00
3300	Trenches, up to 2' deep				
3320	Normal soil	C.Y.	47.00		47.00
3340	Sand or gravel	"	41.75		41.75
3360	Medium clay	"	63.00		63.00

		UNIT	LABOR	MAT.	TOTAL
02220.60	**TRENCHING, Cont'd...**				
3380	Heavy clay	C.Y.	94.00		94.00
3390	Loose rock	"	130		130
3400	Trenches, to 6' deep				
3420	Normal soil	C.Y.	54.00		54.00
3440	Sand or gravel	"	47.00		47.00
3460	Medium clay	"	75.00		75.00
3480	Heavy clay	"	130		130
3500	Loose rock	"	190		190
3590	Backfill trenches				
3600	With compaction				
3620	By hand	C.Y.	31.25		31.25
3640	By 60 hp tracked dozer	"	2.05		2.05
02220.70	**ROADWAY EXCAVATION**				
0100	Roadway excavation				
0110	1/4 mile haul	C.Y.	2.26		2.26
0120	2 mile haul	"	3.77		3.77
0130	5 mile haul	"	5.66		5.66
3000	Spread base course	"	2.83		2.83
3100	Roll and compact	"	3.77		3.77
02220.71	**BASE COURSE**				
1019	Base course, crushed stone				
1020	3" thick	S.Y.	0.56	3.15	3.71
1030	4" thick	"	0.61	4.14	4.75
1040	6" thick	"	0.66	6.30	6.96
2500	Base course, bank run gravel				
3020	4" deep	S.Y.	0.59	2.28	2.87
3040	6" deep	"	0.64	3.32	3.96
4000	Prepare and roll sub base				
4020	Minimum	S.Y.	0.56		0.56
4030	Average	"	0.70		0.70
4040	Maximum	"	0.94		0.94
02220.90	**HAND EXCAVATION**				
0980	Excavation				
1000	To 2' deep				
1020	Normal soil	C.Y.	41.75		41.75
1040	Sand and gravel	"	37.50		37.50
1060	Medium clay	"	47.00		47.00
1080	Heavy clay	"	54.00		54.00
1100	Loose rock	"	63.00		63.00
1200	To 6' deep				
1220	Normal soil	C.Y.	54.00		54.00
1240	Sand and gravel	"	47.00		47.00
1260	Medium clay	"	63.00		63.00
1280	Heavy clay	"	75.00		75.00
1300	Loose rock	"	94.00		94.00
2020	Backfilling foundation without compaction, 6" lifts	"	23.50		23.50
2200	Compaction of backfill around structures or in trench				
2220	By hand with air tamper	C.Y.	26.75		26.75
2240	By hand with vibrating plate tamper	"	25.00		25.00
2250	1 ton roller	"	41.00		41.00
5400	Miscellaneous hand labor				

		UNIT	LABOR	MAT.	TOTAL
02220.90	**HAND EXCAVATION, Cont'd...**				
5440	Trim slopes, sides of excavation	S.F.	0.06		0.06
5450	Trim bottom of excavation	"	0.07		0.07
5460	Excavation around obstructions and services	C.Y.	130		130
02240.05	**SOIL STABILIZATION**				
0100	Straw bale secured with rebar	L.F.	1.25	1.82	3.07
0120	Filter barrier, 18" high filter fabric	"	3.75	1.82	5.57
0130	Sediment fence, 36" fabric with 6" mesh	"	4.69	4.32	9.01
1000	Soil stabilization with tar paper, burlap, straw and stakes	S.F.	0.05	0.36	0.41
02270.40	**RIPRAP**				
0100	Riprap				
0110	Crushed stone blanket, max size 2-1/2"	TON	62.00	30.50	92.50
0120	Stone, quarry run, 300 lb. stones	"	57.00	38.25	95.25
0130	400 lb. stones	"	53.00	38.25	91.25
0140	500 lb. stones	"	49.25	38.25	87.50
0150	750 lb. stones	"	46.25	38.25	84.50
0160	Dry concrete riprap in bags 3" thick, 80 lb. per bag	BAG	3.08	5.95	9.03
02280.20	**SOIL TREATMENT**				
1100	Soil treatment, termite control pretreatment				
1120	Under slabs	S.F.	0.20	0.17	0.37
1140	By walls	"	0.25	0.17	0.42
02290.30	**WEED CONTROL**				
1000	Weed control, bromicil, 15 lb./acre, wettable powder	ACRE	190	290	480
1100	Vegetation control, by application of plant killer	S.Y.	0.15	0.02	0.17
1200	Weed killer, lawns and fields	"	0.07	0.24	0.31
02360.50	**PRESTRESSED PILING**				
0980	Prestressed concrete piling, less than 60' long				
1000	10" sq.	L.F.	5.28	11.25	16.53
1002	12" sq.	"	5.51	15.75	21.26
1480	Straight cylinder, less than 60' long				
1500	12" dia.	L.F.	5.76	14.75	20.51
1540	14" dia.	"	5.90	19.75	25.65
02360.60	**STEEL PILES**				
1000	H-section piles				
1010	8x8				
1020	36 lb/ft				
1021	30' long	L.F.	10.50	16.50	27.00
1022	40' long	"	8.46	16.50	24.96
5000	Tapered friction piles, with fluted steel casing, up to 50'				
5002	With 4000 psi concrete no reinforcing				
5040	12" dia.	L.F.	6.34	20.00	26.34
5060	14" dia.	"	6.50	23.00	29.50
02360.65	**STEEL PIPE PILES**				
1000	Concrete filled, 3000# concrete, up to 40'				
1100	8" dia.	L.F.	9.06	19.75	28.81
1120	10" dia.	"	9.40	24.25	33.65
1140	12" dia.	"	9.76	29.50	39.26
2000	Pipe piles, non-filled				
2020	8" dia.	L.F.	7.05	15.50	22.55
2040	10" dia.	"	7.25	18.25	25.50
2060	12" dia.	"	7.46	22.50	29.96

DIVISION # 02 SITEWORK

		UNIT	LABOR	MAT.	TOTAL
02360.65	**STEEL PIPE PILES, Cont'd...**				
2520	Splice				
2540	8" dia.	EA.	75.00	53.00	128
2560	10" dia.	"	75.00	56.00	131
2580	12" dia.	"	94.00	63.00	157
2680	Standard point				
2700	8" dia.	EA.	75.00	68.00	143
2740	10" dia.	"	75.00	77.00	152
2760	12" dia.	"	94.00	120	214
2880	Heavy duty point				
2900	8" dia.	EA.	94.00	85.00	179
2920	10" dia.	"	94.00	100	194
2940	12" dia.	"	130	140	270
02360.80	**WOOD AND TIMBER PILES**				
0080	Treated wood piles, 12" butt, 8" tip				
0100	25' long	L.F.	12.75	8.94	21.69
0110	30' long	"	10.50	9.55	20.05
0120	35' long	"	9.06	9.55	18.61
0125	40' long	"	7.93	9.55	17.48
02510.20	**ASPHALT SURFACES**				
0050	Asphalt wearing surface, for flexible pavement				
0100	1" thick	S.Y.	2.11	4.90	7.01
0120	1-1/2" thick	"	2.53	7.51	10.04
1000	Binder course				
1010	1-1/2" thick	S.Y.	2.35	6.43	8.78
1030	2" thick	"	2.88	8.58	11.46
2000	Bituminous sidewalk, no base				
2020	2" thick	S.Y.	2.48	10.75	13.23
2040	3" thick	"	2.64	16.00	18.64
02520.10	**CONCRETE PAVING**				
1080	Concrete paving, reinforced, 5000 psi concrete				
2000	6" thick	S.Y.	19.75	25.00	44.75
2005	7" thick	"	21.25	27.50	48.75
2010	8" thick	"	22.75	31.00	53.75
02605.30	**MANHOLES**				
0100	Precast sections, 48" dia.				
0110	Base section	EA.	180	190	370
0120	1'0" riser	"	140	53.00	193
0130	1'4" riser	"	150	64.00	214
0140	2'8" riser	"	160	96.00	256
0150	4'0" riser	"	180	180	360
0160	2'8" cone top	"	210	120	330
0170	Precast manholes, 48" dia.				
0180	4' deep	EA.	420	430	850
0200	6' deep	"	530	660	1,190
0250	7' deep	"	600	760	1,360
0260	8' deep	"	700	850	1,550
0280	10' deep	"	850	950	1,800
1000	Cast-in-place, 48" dia., with frame and cover				
1100	5' deep	EA.	1,060	570	1,630
1120	6' deep	"	1,210	750	1,960
1140	8' deep	"	1,410	1,090	2,500

		UNIT	LABOR	MAT.	TOTAL
02605.30	**MANHOLES, Cont'd...**				
1160	10' deep	EA.	1,690	1,270	2,960
1480	Brick manholes, 48" dia. with cover, 8" thick				
1500	4' deep	EA.	470	670	1,140
1501	6' deep	"	520	840	1,360
1505	8' deep	"	580	1,080	1,660
1510	10' deep	"	670	1,350	2,020
4200	Frames and covers, 24" diameter				
4210	300 lb	EA.	37.50	390	428
4220	400 lb	"	41.75	410	452
4980	Steps for manholes				
5000	7" x 9"	EA.	7.50	13.25	20.75
5020	8" x 9"	"	8.33	16.75	25.08
02610.40	**DUCTILE IRON PIPE**				
0990	Ductile iron pipe, cement lined, slip-on joints				
1000	4"	L.F.	5.87	14.75	20.62
1010	6"	"	6.21	18.00	24.21
1020	8"	"	6.60	23.50	30.10
1190	Mechanical joint pipe				
1200	4"	L.F.	8.12	17.75	25.87
1210	6"	"	8.80	21.25	30.05
1220	8"	"	9.60	27.75	37.35
1480	Fittings, mechanical joint				
1500	90 degree elbow				
1520	4"	EA.	25.00	220	245
1540	6"	"	28.75	290	319
1560	8"	"	37.50	410	448
1700	45 degree elbow				
1720	4"	EA.	25.00	230	255
1740	6"	"	28.75	260	289
1760	8"	"	37.50	360	398
02610.60	**PLASTIC PIPE**				
0110	PVC, class 150 pipe				
0120	4" dia.	L.F.	5.28	7.07	12.35
0130	6" dia.	"	5.71	13.25	18.96
0140	8" dia.	"	6.03	21.25	27.28
0165	Schedule 40 pipe				
0170	1-1/2" dia.	L.F.	2.20	3.92	6.12
0180	2" dia.	"	2.34	4.60	6.94
0185	2-1/2" dia.	"	2.50	5.11	7.61
0190	3" dia.	"	2.68	7.36	10.04
0200	4" dia.	"	3.12	10.50	13.62
0210	6" dia.	"	3.75	17.75	21.50
0240	90 degree elbows				
0250	1"	EA.	6.25	2.63	8.88
0260	1-1/2"	"	6.25	4.44	10.69
0270	2"	"	6.82	4.99	11.81
0280	2-1/2"	"	7.50	12.50	20.00
0290	3"	"	8.33	15.25	23.58
0300	4"	"	9.38	42.00	51.38
0310	6"	"	12.50	94.00	107
0500	Couplings				

		UNIT	LABOR	MAT.	TOTAL
02610.60	**PLASTIC PIPE, Cont'd...**				
0510	1"	EA.	6.25	1.56	7.81
0520	1-1/2"	"	6.25	1.82	8.07
0530	2"	"	6.82	3.14	9.96
0540	2-1/2"	"	7.50	6.29	13.79
0550	3"	"	8.33	11.75	20.08
0560	4"	"	9.38	26.25	35.63
0580	6"	"	12.50	34.00	46.50
02610.90	**VITRIFIED CLAY PIPE**				
0100	Vitrified clay pipe, extra strength				
1020	6" dia.	L.F.	9.60	4.34	13.94
1040	8" dia.	"	10.00	5.20	15.20
1050	10" dia.	"	10.50	7.97	18.47
02670.10	**WELLS**				
0980	Domestic water, drilled and cased				
1000	4" dia.	L.F.	63.00	17.50	80.50
1020	6" dia.	"	71.00	19.25	90.25
02690.10	**STORAGE TANKS**				
0080	Oil storage tank, underground				
0090	Steel				
1000	500 gals	EA.	260	850	1,110
1020	1,000 gals	"	350	1,480	1,830
1980	Fiberglass, double wall				
2000	550 gals	EA.	350	2,780	3,130
2020	1,000 gals	"	350	3,590	3,940
2520	Above ground				
2530	Steel				
2540	275 gals	EA.	210	350	560
2560	500 gals	"	350	1,110	1,460
2570	1,000 gals	"	420	1,660	2,080
3020	Fill cap	"	53.00	46.00	99.00
3040	Vent cap	"	53.00	46.00	99.00
3100	Level indicator	"	53.00	110	163
02720.70	**UNDERDRAIN**				
1480	Drain tile, clay				
1500	6" pipe	L.F.	4.69	4.11	8.80
1520	8" pipe	"	4.91	6.55	11.46
1580	Porous concrete, standard strength				
1600	6" pipe	L.F.	4.69	3.90	8.59
1620	8" pipe	"	4.91	5.69	10.60
1800	Corrugated metal pipe, perforated type				
1810	6" pipe	L.F.	5.28	7.04	12.32
1820	8" pipe	"	5.56	8.33	13.89
1980	Perforated clay pipe				
2000	6" pipe	L.F.	6.03	4.95	10.98
2020	8" pipe	"	6.21	6.63	12.84
2480	Drain tile, concrete				
2500	6" pipe	L.F.	4.69	3.23	7.92
2520	8" pipe	"	4.91	5.02	9.93
4980	Perforated rigid PVC underdrain pipe				
5000	4" pipe	L.F.	3.52	1.79	5.31
5100	6" pipe	"	4.22	3.29	7.51

DIVISION # 02 SITEWORK

		UNIT	LABOR	MAT.	TOTAL
02720.70	**UNDERDRAIN, Cont'd...**				
5150	8" pipe	L.F.	4.69	4.66	9.35
6980	Underslab drainage, crushed stone				
7000	3" thick	S.F.	0.70	0.30	1.00
7120	4" thick	"	0.81	0.37	1.18
7140	6" thick	"	0.88	0.39	1.27
7180	Plastic filter fabric for drain lines	"	0.37	0.15	0.52
02730.10	**SANITARY SEWERS**				
0980	Clay				
1000	6" pipe	L.F.	7.04	4.95	11.99
2980	PVC				
3000	4" pipe	L.F.	5.28	2.49	7.77
3010	6" pipe	"	5.56	4.99	10.55
02740.10	**DRAINAGE FIELDS**				
0080	Perforated PVC pipe, for drain field				
0100	4" pipe	L.F.	4.69	1.88	6.57
0120	6" pipe	"	5.03	3.53	8.56
02740.50	**SEPTIC TANKS**				
0980	Septic tank, precast concrete				
1000	1000 gals	EA.	350	750	1,100
1200	2000 gals	"	530	1,410	1,940
1310	Leaching pit, precast concrete, 72" diameter				
1320	3' deep	EA.	260	550	810
1340	6' deep	"	300	690	990
1360	8' deep	"	350	860	1,210
02810.40	**LAWN IRRIGATION**				
0480	Residential system, complete				
0500	Minimum	ACRE			19,920
0520	Maximum	"			37,920
02830.10	**CHAIN LINK FENCE**				
0230	Chain link fence, 9 ga., galvanized, with posts 10' o.c.				
0250	4' high	L.F.	2.68	6.70	9.38
0260	5' high	"	3.41	8.98	12.39
0270	6' high	"	4.69	10.25	14.94
1161	Fabric, galvanized chain link, 2" mesh, 9 ga.				
1163	4' high	L.F.	1.25	3.05	4.30
1164	5' high	"	1.50	3.71	5.21
1165	6' high	"	1.87	3.91	5.78
02830.20	**WOOD FENCE**				
0010	4X4 posts w/2x4 horizontals, 4'-8' high				
0020	Cedar, 1x4 planks, picket	S.F.	0.93	0.85	1.78
0030	1x6 planks, privacy	"	0.75	2.79	3.54
0040	1x8 planks, privacy	"	0.68	2.55	3.23
0050	Redwood, 1x4 planks, picket	"	0.93	0.71	1.64
0060	1x6 planks, privacy	"	0.75	2.65	3.40
0070	1x8 planks	"	0.68	2.31	2.99
0080	Treated pine, 1x4 planks, picket	"	0.93	1.40	2.33
0100	1x6 planks, privacy	"	0.75	1.37	2.12
0110	1x8 planks, privacy	"	0.68	1.35	2.03
0210	4' high				
0220	Composite, 1x4 planks, picket	S.F.	1.17	0.77	1.94
0230	1x6 planks, privacy	"	0.93	2.65	3.58

DIVISION # 02 SITEWORK

		UNIT	LABOR	MAT.	TOTAL
02830.20	**WOOD FENCE, Cont'd...**				
0240	1x8 planks	S.F.	0.78	2.31	3.09
0250	Vinyl, 1x4 planks, picket	"	1.17	1.37	2.54
0260	1x6 planks, privacy	"	0.93	2.44	3.37
0270	1x8 planks, privacy	"	0.78	2.39	3.17
0280	Gate, cedar or redwood, 3' w, 1x4 planks, picket	EA.	31.25	24.25	55.50
0290	1x6 planks, privacy	"	37.50	66.00	104
0300	1x8 planks, privacy	"	41.75	76.00	118
02830.70	**RECREATIONAL COURTS**				
1000	Walls, galvanized steel				
1020	8' high	L.F.	7.50	13.75	21.25
1040	10' high	"	8.33	16.25	24.58
1060	12' high	"	9.87	18.75	28.62
1200	Vinyl coated				
1220	8' high	L.F.	7.50	13.25	20.75
1240	10' high	"	8.33	16.25	24.58
1260	12' high	"	9.87	18.00	27.87
2010	Gates, galvanized steel				
2200	Single, 3' transom				
2210	3'x7'	EA.	190	320	510
2220	4'x7'	"	210	340	550
2230	5'x7'	"	250	470	720
2240	6'x7'	"	300	510	810
2400	Vinyl coated				
2405	Single, 3' transom				
2410	3'x7'	EA.	190	630	820
2420	4'x7'	"	210	690	900
2430	5'x7'	"	250	690	940
2440	6'x7'	"	300	710	1,010
02840.40	**PARKING BARRIERS**				
3010	Bollard, conc. filled, 8' long				
3020	6" dia.	EA.	32.25	410	442
3030	8" dia.	"	40.25	600	640
3040	12" dia.	"	48.25	780	828
02910.10	**SHRUB & TREE MAINTENANCE**				
1000	Moving shrubs on site				
1220	3' high	EA.	37.50		37.50
1240	4' high	"	41.75		41.75
2000	Moving trees on site				
3060	6' high	EA.	47.00		47.00
3080	8' high	"	53.00		53.00
3100	10' high	"	70.00		70.00
3110	Palm trees				
3140	10' high	EA.	70.00		70.00
3144	40' high	"	420		420
02920.10	**TOPSOIL**				
0005	Spread topsoil, with equipment				
0010	Minimum	C.Y.	11.25		11.25
0020	Maximum	"	14.25		14.25
0080	By hand				
0100	Minimum	C.Y.	37.50		37.50
0110	Maximum	"	47.00		47.00

		UNIT	LABOR	MAT.	TOTAL
02920.10	**TOPSOIL, Cont'd...**				
0980	Area preparation for seeding (grade, rake and clean)				
1000	Square yard	S.Y.	0.30		0.30
1020	By acre	ACRE	1,500		1,500
2000	Remove topsoil and stockpile on site				
2020	4" deep	C.Y.	9.44		9.44
2040	6" deep	"	8.71		8.71
2200	Spreading topsoil from stock pile				
2220	By loader	C.Y.	10.25		10.25
2240	By hand	"	110		110
2260	Top dress by hand	S.Y.	1.13		1.13
2280	Place imported top soil				
2300	By loader				
2320	4" deep	S.Y.	1.13		1.13
2340	6" deep	"	1.25		1.25
2360	By hand				
2370	4" deep	S.Y.	4.16		4.16
2380	6" deep	"	4.69		4.69
5980	Plant bed preparation, 18" deep				
6000	With backhoe/loader	S.Y.	2.83		2.83
6010	By hand	"	6.25		6.25
02930.30	**SEEDING**				
0980	Mechanical seeding, 175 lb/acre				
1000	By square yard	S.Y.	0.09	0.67	0.76
1020	By acre	ACRE	500	2,690	3,190
2040	450 lb/acre				
2060	By square yard	S.Y.	0.12	1.08	1.20
2080	By acre	ACRE	620	4,110	4,730
5980	Seeding by hand, 10 lb per 100 s.y.				
6000	By square yard	S.Y.	0.12	0.71	0.83
6010	By acre	ACRE	630	2,850	3,480
8010	Reseed disturbed areas	S.F.	0.18	0.64	0.82
02970.10	**FERTILIZING**				
0080	Fertilizing (23#/1000 sf)				
0100	By square yard	S.Y.	0.12	0.03	0.15
0120	By acre	ACRE	620	140	760
2980	Liming (70#/1000 sf)				
3000	By square yard	S.Y.	0.16	0.03	0.19
3020	By acre	ACRE	830	130	960
02980.10	**LANDSCAPE ACCESSORIES**				
0100	Steel edging, 3/16" x 4"	L.F.	0.46	0.62	1.08
0200	Landscaping stepping stones, 15"x15", white	EA.	1.87	5.55	7.42
6000	Wood chip mulch	C.Y.	25.00	45.75	70.75
6010	2" thick	S.Y.	0.75	2.81	3.56
6020	4" thick	"	1.07	5.28	6.35
6030	6" thick	"	1.36	7.92	9.28
6200	Gravel mulch, 3/4" stone	C.Y.	37.50	36.25	73.75
6300	White marble chips, 1" deep	S.F.	0.37	0.72	1.09
6980	Peat moss				
7000	2" thick	S.Y.	0.83	3.30	4.13
7020	4" thick	"	1.25	6.34	7.59
7030	6" thick	"	1.56	9.67	11.23

		UNIT	LABOR	MAT.	TOTAL
02980.10	**LANDSCAPE ACCESSORIES, Cont'd...**				
7980	Landscaping timbers, treated lumber				
8000	4" x 4"	L.F.	1.25	1.29	2.54
8020	6" x 6"	"	1.34	2.57	3.91
8040	8" x 8"	"	1.56	4.18	5.74

		UNIT	COST
02999.10	**DEMOLITION**		
3200	Selective Building Removals		
3210	No Cutting or Disposal Included		
3220	Concrete - Hand Work		
3230	8" Walls - Reinforced	S.F.	13.25
3240	Non - Reinforced	"	8.80
3250	12" Walls - Reinforced	"	21.25
3260	12" Footings x 24" wide	L.F.	19.25
3270	x 36" wide	"	26.50
3280	16" Footings x 24" wide	S.F.	35.25
3290	6" Structural Slab - Reinforced	"	8.12
3300	8" Structural Slab - Reinforced	"	10.00
3310	4" Slab on Ground - Reinforced	"	3.77
3320	Non - Reinforced	"	2.81
3330	6" Slab on Ground - Reinforced	"	4.49
3340	Non - Reinforced	"	3.35
3350	Stairs - Reinforced	"	14.00
3360	Masonry - Hand Work		
3370	4" Brick or Stone Walls	S.F.	2.54
3380	4" Brick and 8" Backup Block or Tile	"	4.69
3390	4" Block or Tile Partitions	"	2.48
3400	6" Block or Tile Partitions	"	2.70
3410	8" Block or Tile Partitions	"	3.10
3420	12" Block or Tile Partitions	"	3.91
3430	Miscellaneous - Hand Work		
3440	Acoustical Ceilings - Attached	S.F.	0.72
3450	Suspended (Including Grid)	"	0.41
3460	Asbestos - Pipe	L.F.	53.00
3470	Ceilings and Walls	S.F.	17.50
3480	Columns and Beams	"	42.25
3490	Tile Flooring	"	1.97
3500	Cabinets and Tops	L.F.	11.75
3510	Carpet	S.F.	0.36
3520	Ceramic and Quarry Tile	"	1.40
3530	Doors and Frames - Metal	EA.	62.00
3540	Wood	"	53.00
3550	Drywall Ceilings - Attached	S.F.	0.88
3560	Drywall on Wood or Metal Studs - 2 Sides	"	0.98
3570	Paint Removal - Doors and Windows	"	0.88
3580	Walls	"	0.72
3590	Plaster Ceilings - Attached (Including Iron)	"	1.25
3600	on Wood or Metal Studs	"	1.19
3610	Roofing - Builtup	"	1.19
3620	Shingles - Asphalt and Wood	"	0.41
3630	Terrazzo Flooring	"	1.97
3640	Vinyl Flooring	"	0.41
3650	Wall Coverings	"	0.78
3660	Windows	EA.	42.25
3670	Wood Flooring	S.F.	0.41
4300	Site Removals (Including Loading)		
4310	4" Concrete Walks - Labor Only (Non-Reinforced)	S.F.	2.81
4320	Machine Only (Non-Reinforced)	"	0.93
4330	6" Concrete Drives - Labor Only (Non-Reinforced)	"	3.52

		UNIT	COST
02999.10	**DEMOLITION, Cont'd...**		
4340	Machine Only (Reinforced)	S.F.	1.25
4350	6" x 18" Concrete Curb - Machine	L.F.	2.40
4360	Curb and Gutter - Machine	"	3.12
4370	2" Asphalt - Machine	S.F.	0.54
4380	Fencing - 8' Hand	L.F.	3.22
02999.20	**EARTHWORK**		
1000	GRADING - Hand - 4" - Site	S.F.	0.32
1010	Hand – 4" Building	"	0.54
2000	EXCAVATION - Hand - Open - Soft (Sand)	C.Y.	42.25
2010	Medium (Clay)	"	53.00
2020	Hard (Shale)	"	85.00
2030	Add for Trench or Pocket	PCT.	15.00
3000	BACK FILL - Hand - Not Comp. (Site Borrow)	C.Y.	19.25
4000	BACK FILL - Hand - Comp.(Site Borrow)		
4010	12" Lifts - Building - No Machine	C.Y.	32.50
4020	With Machine	"	26.50
4030	18" Lifts - Building - No Machine	"	30.25
4040	With Machine	"	24.75
02999.50	**DRAINAGE**		
4000	BUILDING FOUNDATION DRAINAGE		
4010	4" Clay Pipe	L.F.	8.53
4020	4" Plastic Pipe - Perforated	"	6.88
4030	6" Clay Pipe	"	9.71
4040	6" Plastic Pipe - Perforated	"	10.97
4050	Add for Porous Surround - 2' x 2'	"	12.59
02999.60	**PAVEMENT, CURBS AND WALKS**		
2000	CURBS AND GUTTERS		
2100	Concrete - Cast in Place (Machine Placed)		
2110	Curb - 6" x 12"	L.F.	19.43
2120	6" x 18"	"	22.88
2130	6" x 24"	"	26.22
2140	6" x 30"	"	30.17
2150	Curb and Gutter - 6" x 12"	"	28.33
2160	6" x 18"	"	33.50
2170	6" x 24"	"	38.50
2180	Add for Hand Placed	"	8.80
2190	Add for 2 #5 Reinf. Rods	"	3.37
2191	Add for Curves and Radius Work	PCT.	40.00
2200	Concrete Precast - 6" x 10" x 8"	L.F.	21.12
2210	6" x 9" x 8"	"	19.09
2250	6" x 8"	"	17.07
2300	Bituminous - 6" x 8"	"	25.91
2400	Granite - 6" x 16"	"	40.68
2500	Timbers - Treated - 6" x 6"	"	14.27
2600	Plastic - 6" x 6"	"	18.61
3000	WALKS		
3010	Bituminous - 1 1/2" with 4" Sand Base	S.F.	3.74
3020	2" with 4" Sand Base	"	4.14
3030	Concrete - 4" - Broom Finish	"	8.92
3040	5" - Broom Finish	"	9.43
3050	6" - Broom Finish	"	10.08

		UNIT	COST
02999.60	**PAVEMENT, CURBS AND WALKS, Cont'd...**		
3060	Add for 6" x 6", 10 - 10 Mesh	S.F.	1.64
3070	Add for 4" Sand Base	"	1.38
3080	Add for Exposed Aggregate	"	2.17
3090	Crushed Rock - 4"	"	1.48
4000	Brick - 4" - with 2" Sand Cushion	"	21.62
4010	4" - with 2" Mortar Setting Bed	"	27.00
4020	Flagstone – 1 1/4" - with 4" Sand Cushion	"	38.25
4030	1¼" - with 2" Mortar Setting Bed	"	41.50
4040	Precast Block - 1" Colored with 4" Sand Cushion	"	7.70
4050	2" Colored with 4" Sand Cushion	"	8.93
4060	Wood - 2" Boards on 6" x 6" Timbers	"	13.63
4070	2" Boards on 4" x 4" Timbers	"	11.47
4080	Slate - 1 1/4" - with 2" Mortar Setting Bed	"	48.25

Design & Construction Resources

TABLE OF CONTENTS

PAGE

03100.03 - FORMWORK ACCESSORIES	03-2
03110.05 - BEAM FORMWORK	03-2
03110.15 - COLUMN FORMWORK	03-3
03110.18 - CURB FORMWORK	03-3
03110.20 - ELEVATED SLAB FORMWORK	03-3
03110.25 - EQUIPMENT PAD FORMWORK	03-3
03110.35 - FOOTING FORMWORK	03-3
03110.50 - GRADE BEAM FORMWORK	03-4
03110.53 - PILE CAP FORMWORK	03-4
03110.55 - SLAB/MAT FORMWORK	03-4
03110.60 - STAIR FORMWORK	03-4
03110.65 - WALL FORMWORK	03-4
03110.70 - INSULATED CONCRETE FORMS	03-5
03110.90 - MISCELLANEOUS FORMWORK	03-5
03210.05 - BEAM REINFORCING	03-6
03210.15 - COLUMN REINFORCING	03-6
03210.20 - ELEVATED SLAB REINFORCING	03-6
03210.25 - EQUIP. PAD REINFORCING	03-6
03210.35 - FOOTING REINFORCING	03-6
03210.45 - FOUNDATION REINFORCING	03-7
03210.50 - GRADE BEAM REINFORCING	03-7
03210.53 - PILE CAP REINFORCING	03-7
03210.55 - SLAB/MAT REINFORCING	03-7
03210.60 - STAIR REINFORCING	03-8
03210.65 - WALL REINFORCING	03-8
03250.40 - CONCRETE ACCESSORIES	03-8
03300.10 - CONCRETE ADMIXTURES	03-8
03350.10 - CONCRETE FINISHES	03-8
03360.10 - PNEUMATIC CONCRETE	03-9
03370.10 - CURING CONCRETE	03-9
03380.05 - BEAM CONCRETE	03-9
03380.15 - COLUMN CONCRETE	03-9
03380.25 - EQUIPMENT PAD CONCRETE	03-10
03380.35 - FOOTING CONCRETE	03-10
03380.50 - GRADE BEAM CONCRETE	03-10
03380.53 - PILE CAP CONCRETE	03-10
03380.55 - SLAB/MAT CONCRETE	03-11
03380.58 - SIDEWALKS	03-11
03380.60 - STAIR CONCRETE	03-11
03380.65 - WALL CONCRETE	03-11
03730.10 - CONCRETE REPAIR	03-12
QUICK ESTIMATING	03A-14

		UNIT	LABOR	MAT.	TOTAL
03100.03	**FORMWORK ACCESSORIES**				
1000	Column clamps				
1010	Small, adjustable, 24"x24"	EA.			130
1020	Medium 36"x36"	"			280
1030	Large 60"x60"	"			440
2000	Forming hangers				
2010	Iron 14 ga.	EA.			1.58
2020	22 ga.	"			1.58
3000	Snap ties				
3010	Short-end with washers, 6" long	EA.			5.11
3020	12" long	"			6.00
4000	18" long	"			6.53
4010	24" long	"			7.10
4020	Long-end with washers, 6' long	"			5.44
4030	12" long	"			6.54
4040	18" long	"			7.10
4050	24" long	"			8.08
5000	Stakes				
5010	Round, pre-drilled holes, 12" long	EA.			5.30
5020	18" long	"			5.79
5030	24" long	"			7.33
5040	30" long	"			9.11
5050	36" long	"			9.49
5060	48" long	"			17.50
5070	I beam type, 12" long	"			8.80
6000	18" long	"			9.29
6010	24" long	"			10.50
6020	30" long	"			14.25
6030	36" long	"			17.50
7000	48" long	"			21.00
7010	Taper ties				
7020	50K, 1-1/4" to 1", 35" long	EA.			66.00
8000	45" long	"			77.00
8010	55" long	"			88.00
8020	Walers				
8030	5" deep, 4' long	EA.			230
8040	8' long	"			360
8050	12' long	"			510
9000	16' long	"			660
9010	8" deep, 4' long	"			410
9020	8' long	"			350
9030	12' long	"			930
9040	16' long	"			1,190
03110.05	**BEAM FORMWORK**				
1000	Beam forms, job built				
1020	Beam bottoms				
1040	1 use	S.F.	8.04	4.32	12.36
1080	3 uses	"	7.42	1.94	9.36
1120	5 uses	"	6.89	1.46	8.35
2000	Beam sides				
2020	1 use	S.F.	5.36	3.09	8.45
2060	3 uses	"	4.82	1.60	6.42

		UNIT	LABOR	MAT.	TOTAL
03110.05	**BEAM FORMWORK, Cont'd...**				
2100	5 uses	S.F.	4.38	1.31	5.69
03110.15	**COLUMN FORMWORK**				
1000	Column, square forms, job built				
1020	8" x 8" columns				
1040	1 use	S.F.	9.65	3.39	13.04
1080	3 uses	"	8.94	1.55	10.49
1120	5 uses	"	8.32	1.20	9.52
1300	16" x 16" columns				
1320	1 use	S.F.	8.04	2.95	10.99
1360	3 uses	"	7.54	1.24	8.78
1390	5 uses	"	7.10	0.92	8.02
2000	Round fiber forms, 1 use				
2040	10" dia.	L.F.	9.65	4.08	13.73
2120	18" dia.	"	11.50	14.25	25.75
2180	36" dia.	"	14.75	32.25	47.00
03110.18	**CURB FORMWORK**				
0980	Curb forms				
0990	Straight, 6" high				
1000	1 use	L.F.	4.82	1.98	6.80
1040	3 uses	"	4.38	0.88	5.26
1080	5 uses	"	4.02	0.72	4.74
1090	Curved, 6" high				
2000	1 use	L.F.	6.03	2.13	8.16
2040	3 uses	"	5.36	1.02	6.38
2080	5 uses	"	4.92	0.88	5.80
03110.20	**ELEVATED SLAB FORMWORK**				
0100	Elevated slab formwork				
1000	Slab, with drop panels				
1020	1 use	S.F.	3.86	3.63	7.49
1060	3 uses	"	3.57	1.62	5.19
1100	5 uses	"	3.32	1.29	4.61
2000	Floor slab, hung from steel beams				
2020	1 use	S.F.	3.71	2.92	6.63
2060	3 uses	"	3.44	1.45	4.89
2100	5 uses	"	3.21	1.08	4.29
3000	Floor slab, with pans or domes				
3020	1 use	S.F.	4.38	5.20	9.58
3060	3 uses	"	4.02	3.14	7.16
3100	5 uses	"	3.71	2.61	6.32
9030	Equipment curbs, 12" high				
9035	1 use	L.F.	4.82	2.64	7.46
9060	3 uses	"	4.38	1.47	5.85
9100	5 uses	"	4.02	1.11	5.13
03110.25	**EQUIPMENT PAD FORMWORK**				
1000	Equipment pad, job built				
1020	1 use	S.F.	6.03	3.46	9.49
1040	2 uses	"	5.68	2.07	7.75
1060	3 uses	"	5.36	1.66	7.02

		UNIT	LABOR	MAT.	TOTAL
03110.35	**FOOTING FORMWORK**				
2000	Wall footings, job built, continuous				
2040	1 use	S.F.	4.82	1.62	6.44
2060	3 uses	"	4.38	0.94	5.32
2090	5 uses	"	4.02	0.72	4.74
3000	Column footings, spread				
3020	1 use	S.F.	6.03	1.71	7.74
3060	3 uses	"	5.36	0.91	6.27
3100	5 uses	"	4.82	0.69	5.51
03110.50	**GRADE BEAM FORMWORK**				
1000	Grade beams, job built				
1020	1 use	S.F.	4.82	2.55	7.37
1060	3 uses	"	4.38	1.12	5.50
1100	5 uses	"	4.02	0.77	4.79
03110.53	**PILE CAP FORMWORK**				
1500	Pile cap forms, job built				
1510	Square				
1520	1 use	S.F.	6.03	2.91	8.94
1560	3 uses	"	5.36	1.33	6.69
1600	5 uses	"	4.82	0.97	5.79
03110.55	**SLAB/MAT FORMWORK**				
3000	Mat foundations, job built				
3020	1 use	S.F.	6.03	2.54	8.57
3060	3 uses	"	5.36	1.07	6.43
3100	5 uses	"	4.82	0.72	5.54
3980	Edge forms				
3990	6" high				
4000	1 use	L.F.	4.38	2.54	6.92
4002	3 uses	"	4.02	1.07	5.09
4004	5 uses	"	3.71	0.72	4.43
4014	5 uses	"	4.02	0.68	4.70
5000	Formwork for openings				
5020	1 use	S.F.	9.65	3.47	13.12
5060	3 uses	"	8.04	1.65	9.69
5100	5 uses	"	6.89	1.06	7.95
03110.60	**STAIR FORMWORK**				
1000	Stairway forms, job built				
1020	1 use	S.F.	9.65	4.33	13.98
1040	3 uses	"	8.04	1.88	9.92
1060	5 uses	"	6.89	1.45	8.34
03110.65	**WALL FORMWORK**				
2980	Wall forms, exterior, job built				
3000	Up to 8' high wall				
3120	1 use	S.F.	4.82	2.01	7.83
3160	3 uses	"	4.38	1.37	5.75
3190	5 uses	"	4.02	1.02	5.04
3200	Over 8' high wall				
3220	1 use	S.F.	6.03	3.09	9.12
3240	3 uses	"	5.36	1.60	6.96
3290	5 uses	"	4.82	1.27	6.09
5000	Column pier and pilaster				
5020	1 use	S.F.	9.65	3.09	12.74

		UNIT	LABOR	MAT.	TOTAL
03110.65	**WALL FORMWORK, Cont'd...**				
5060	3 uses	S.F.	8.04	1.70	9.74
5090	5 uses	"	6.89	1.39	8.28
6980	Interior wall forms				
7000	Up to 8' high				
7020	1 use	S.F.	4.38	2.81	7.19
7060	3 uses	"	4.02	1.39	5.41
7100	5 uses	"	3.71	1.01	4.72
7200	Over 8' high				
7220	1 use	S.F.	5.36	3.09	8.45
7260	3 uses	"	4.82	1.60	6.42
7290	5 uses	"	4.38	1.28	5.66
9000	PVC form liner, per side, smooth finish				
9010	1 use	S.F.	4.02	6.46	10.48
9030	3 uses	"	3.71	2.95	6.66
9050	5 uses	"	3.21	1.84	5.05
03110.70	**INSULATED CONCRETE FORMS**				
0010	4" thick, straight	S.F.			4.25
0020	90° corner	"			4.20
0030	45° angle	"			4.56
1000	6" thick, straight	"			4.28
1010	90° corner	"			4.30
1020	45° angle	"			5.00
1030	6" thick, corbel ledge	"			4.94
1040	T-block	"			5.32
2000	8" thick, straight	"			4.46
2010	90° corner	"			4.24
2020	45° angle	"			4.86
2030	corbel ledge	"			5.12
2040	T-block	"			5.53
03110.90	**MISCELLANEOUS FORMWORK**				
1200	Keyway forms (5 uses)				
1220	2 x 4	L.F.	2.41	0.18	2.59
1240	2 x 6	"	2.68	0.27	2.95
1500	Bulkheads				
1510	Walls, with keyways				
1515	2 piece	L.F.	4.38	3.16	7.54
1520	3 piece	"	4.82	3.90	8.72
1560	Elevated slab, with keyway				
1570	2 piece	L.F.	4.02	3.57	7.59
1580	3 piece	"	4.38	5.19	9.57
1600	Ground slab, with keyway				
1620	2 piece	L.F.	3.44	3.90	7.34
1640	3 piece	"	3.71	5.19	8.90
2000	Chamfer strips				
2020	Wood				
2040	1/2" wide	L.F.	1.07	0.23	1.30
2060	3/4" wide	"	1.07	0.30	1.37
2070	1" wide	"	1.07	0.40	1.47
2100	PVC				
2120	1/2" wide	L.F.	1.07	0.86	1.93
2140	3/4" wide	"	1.07	0.95	2.02

		UNIT	LABOR	MAT.	TOTAL
03110.90	**MISCELLANEOUS FORMWORK, Cont'd...**				
2160	1" wide	L.F.	1.07	1.23	2.30
2170	Radius				
2180	1"	L.F.	1.14	1.04	2.18
2200	1-1/2"	"	1.14	1.85	2.99
3000	Reglets				
3020	Galvanized steel, 24 ga.	L.F.	1.93	1.29	3.22
5000	Metal formwork				
5020	Straight edge forms				
5080	8" high	L.F.	3.44	0.23	3.67
5300	Curb form, S-shape				
5310	12" x				
5340	2'	L.F.	6.43	0.43	6.86
5380	3'	"	5.36	0.56	5.92
03210.05	**BEAM REINFORCING**				
0980	Beam-girders				
1000	#3 - #4	TON	1,340	1,220	2,560
1011	#7 - #8	"	900	1,020	1,920
1018	Galvanized				
1020	#3 - #4	TON	1,340	2,080	3,420
1031	#7 - #8	"	900	1,890	2,790
2000	Bond Beams				
2100	#3 - #4	TON	1,790	1,220	3,010
2120	#7 - #8	"	1,190	1,020	2,210
2200	Galvanized				
2210	#3 - #4	TON	1,790	1,990	3,780
2230	#7 - #8	"	1,190	1,890	3,080
03210.15	**COLUMN REINFORCING**				
0980	Columns				
1000	#3 - #4	TON	1,530	1,220	2,750
1015	#7 - #8	"	1,070	1,020	2,090
1100	Galvanized				
1200	#3 - #4	TON	1,530	2,080	3,610
1320	#7 - #8	"	1,070	1,890	2,960
03210.20	**ELEVATED SLAB REINFORCING**				
0980	Elevated slab				
1000	#3 - #4	TON	670	1,220	1,890
1040	#7 - #8	"	540	1,020	1,560
1980	Galvanized				
2000	#3 - #4	TON	670	1,990	2,660
2040	#7 - #8	"	540	1,890	2,430
03210.25	**EQUIP. PAD REINFORCING**				
0980	Equipment pad				
1000	#3 - #4	TON	1,070	1,220	2,290
1040	#7 - #8	"	900	1,020	1,920
03210.35	**FOOTING REINFORCING**				
1000	Footings				
1010	Grade 50				
1020	#3 - #4	TON	900	1,220	2,120
1040	#7 - #8	"	670	1,020	1,690
1055	Grade 60				
1060	#3 - #4	TON	900	1,220	2,120

		UNIT	LABOR	MAT.	TOTAL
03210.35	**FOOTING REINFORCING, Cont'd...**				
1074	#7 - #8	TON	670	1,020	1,690
4980	Straight dowels, 24" long				
5000	1" dia. (#8)	EA.	5.37	3.79	9.16
5040	3/4" dia. (#6)	"	5.37	3.42	8.79
5050	5/8" dia. (#5)	"	4.47	2.95	7.42
5060	1/2" dia. (#4)	"	3.83	2.22	6.05
03210.45	**FOUNDATION REINFORCING**				
0980	Foundations				
1000	#3 - #4	TON	900	1,220	2,120
1040	#7 - #8	"	670	1,020	1,690
1380	Galvanized				
1400	#3 - #4	TON	900	2,080	2,980
1420	#7 - #8	"	670	1,890	2,560
03210.50	**GRADE BEAM REINFORCING**				
0980	Grade beams				
1000	#3 - #4	TON	830	1,220	2,050
1040	#7 - #8	"	630	1,020	1,650
1090	Galvanized				
1100	#3 - #4	TON	830	2,080	2,910
1140	#7 - #8	"	630	1,890	2,520
03210.53	**PILE CAP REINFORCING**				
0980	Pile caps				
1000	#3 - #4	TON	1,340	1,220	2,560
1040	#7 - #8	"	1,070	1,020	2,090
1090	Galvanized				
1100	#3 - #4	TON	1,340	2,080	3,420
1140	#7 - #8	"	1,070	1,890	2,960
03210.55	**SLAB/MAT REINFORCING**				
0980	Bars, slabs				
1000	#3 - #4	TON	900	1,220	2,120
1040	#7 - #8	"	670	1,020	1,690
1980	Galvanized				
2000	#3 - #4	TON	900	2,080	2,980
2020	#5 - #6	"	770	1,960	2,730
2040	#7 - #8	"	670	1,890	2,560
5000	Wire mesh, slabs				
5010	Galvanized				
5015	4x4				
5020	W1.4xW1.4	S.F.	0.35	0.30	0.65
5040	W2.0xW2.0	"	0.38	0.39	0.77
5060	W2.9xW2.9	"	0.41	0.55	0.96
5080	W4.0xW4.0	"	0.44	0.81	1.25
5090	6x6				
5100	W1.4xW1.4	S.F.	0.26	0.28	0.54
5120	W2.0xW2.0	"	0.29	0.39	0.68
5140	W2.9xW2.9	"	0.31	0.53	0.84
5160	W4.0xW4.0	"	0.35	0.57	0.92
5170	Standard				
5175	2x2				
5180	W.9xW.9	S.F.	0.35	0.30	0.65
5185	4x4				

		UNIT	LABOR	MAT.	TOTAL
03210.55	**SLAB/MAT REINFORCING, Cont'd...**				
5190	W1.4xW1.4	S.F.	0.35	0.19	0.54
5500	W4.0xW4.0	"	0.44	0.55	0.99
5580	6x6				
5600	W1.4xW1.4	S.F.	0.26	0.12	0.38
6020	W4.0xW4.0	"	0.35	0.37	0.72
03210.60	**STAIR REINFORCING**				
0980	Stairs				
1000	#3 - #4	TON	1,070	1,220	2,290
1020	#5 - #6	"	900	1,070	1,970
1980	Galvanized				
2000	#3 - #4	TON	1,070	2,080	3,150
2020	#5 - #6	"	900	1,960	2,860
03210.65	**WALL REINFORCING**				
0980	Walls				
1000	#3 - #4	TON	770	1,220	1,990
1040	#7 - #8	"	600	1,020	1,620
1980	Galvanized				
2000	#3 - #4	TON	770	2,080	2,850
2040	#7 - #8	"	600	1,890	2,490
8980	Masonry wall (horizontal)				
9000	#3 - #4	TON	2,150	1,220	3,370
9020	#5 - #6	"	1,790	1,070	2,860
9030	Galvanized				
9040	#3 - #4	TON	2,150	2,080	4,230
9060	#5 - #6	"	1,790	1,960	3,750
9180	Masonry wall (vertical)				
9200	#3 - #4	TON	2,690	1,220	3,910
9220	#5 - #6	"	2,150	1,070	3,220
9230	Galvanized				
9240	#3 - #4	TON	2,690	2,080	4,770
9260	#5 - #6	"	2,150	1,960	4,110
03250.40	**CONCRETE ACCESSORIES**				
1000	Expansion joint, poured				
1010	Asphalt				
1020	1/2" x 1"	L.F.	0.75	0.77	1.52
1040	1" x 2"	"	0.81	2.36	3.17
1060	Liquid neoprene, cold applied				
1080	1/2" x 1"	L.F.	0.76	2.66	3.42
1100	1" x 2"	"	0.83	11.00	11.83
03300.10	**CONCRETE ADMIXTURES**				
1000	Concrete admixtures				
1020	Water reducing admixture	GAL			17.00
1040	Set retarder	"			30.25
1060	Air entraining agent	"			10.25
03350.10	**CONCRETE FINISHES**				
0980	Floor finishes				
1000	Broom	S.F.	0.53		0.53
1020	Screed	"	0.46		0.46
1040	Darby	"	0.46		0.46
1060	Steel float	"	0.62		0.62
4000	Wall finishes				

		UNIT	LABOR	MAT.	TOTAL
03350.10	**CONCRETE FINISHES, Cont'd...**				
4020	Burlap rub, with cement paste	S.F.	0.62	0.05	0.67
4160	Break ties and patch holes	"	0.75		0.75
4170	Carborundum				
4180	Dry rub	S.F.	1.25		1.25
4200	Wet rub	"	1.87		1.87
5000	Floor hardeners				
5010	Metallic				
5020	Light service	S.F.	0.46	0.28	0.74
5040	Heavy service	"	0.62	0.92	1.54
5050	Non-metallic				
5060	Light service	S.F.	0.46	0.15	0.61
5080	Heavy service	"	0.62	0.67	1.29
03360.10	**PNEUMATIC CONCRETE**				
0100	Pneumatic applied concrete (gunite)				
1035	2" thick	S.F.	2.64	4.76	7.40
1040	3" thick	"	3.52	5.91	9.43
1060	4" thick	"	4.22	7.06	11.28
1980	Finish surface				
2000	Minimum	S.F.	2.41		2.41
2020	Maximum	"	4.82		4.82
03370.10	**CURING CONCRETE**				
1000	Sprayed membrane				
1010	Slabs	S.F.	0.07	0.05	0.12
1020	Walls	"	0.09	0.07	0.16
1025	Curing paper				
1030	Slabs	S.F.	0.09	0.07	0.16
2000	Walls	"	0.11	0.07	0.18
2010	Burlap				
2020	7.5 oz.	S.F.	0.12	0.06	0.18
2500	12 oz.	"	0.13	0.08	0.21
03380.05	**BEAM CONCRETE**				
0960	Beams and girders				
0980	2500# or 3000# concrete				
1010	By pump	C.Y.	69.00	98.00	167
1020	By hand buggy	"	37.50	98.00	136
4000	5000# concrete				
4020	By pump	C.Y.	69.00	110	179
4040	By hand buggy	"	37.50	110	148
9460	Bond beam, 3000# concrete				
9470	By pump				
9480	8" high				
9500	4" wide	L.F.	1.52	0.26	1.78
9520	6" wide	"	1.72	0.65	2.37
9530	8" wide	"	1.90	0.83	2.73
9540	10" wide	"	2.11	1.10	3.21
9550	12" wide	"	2.37	1.48	3.85
03380.15	**COLUMN CONCRETE**				
0980	Columns				
0990	2500# or 3000# concrete				
1010	By pump	C.Y.	63.00	98.00	161
3980	5000# concrete				

		UNIT	LABOR	MAT.	TOTAL
03380.15	**COLUMN CONCRETE, Cont'd...**				
4020	By pump	C.Y.	63.00	110	173
03380.25	**EQUIPMENT PAD CONCRETE**				
0960	Equipment pad				
0980	2500# or 3000# concrete				
1000	By chute	C.Y.	12.50	98.00	111
1020	By pump	"	54.00	98.00	152
1050	3500# or 4000# concrete				
1060	By chute	C.Y.	12.50	100	113
1080	By pump	"	54.00	100	154
1110	5000# concrete				
1120	By chute	C.Y.	12.50	110	123
1140	By pump	"	54.00	110	164
03380.35	**FOOTING CONCRETE**				
0980	Continuous footing				
0990	2500# or 3000# concrete				
1000	By chute	C.Y.	12.50	98.00	111
1010	By pump	"	47.50	98.00	146
4000	5000# concrete				
4010	By chute	C.Y.	12.50	110	123
4020	By pump	"	47.50	110	158
4980	Spread footing				
5000	2500# or 3000# concrete				
5010	Under 5 cy				
5020	By chute	C.Y.	12.50	98.00	111
5040	By pump	"	51.00	98.00	149
7200	5000# concrete				
7205	Under 5 c.y.				
7210	By chute	C.Y.	12.50	110	123
7220	By pump	"	51.00	110	161
03380.50	**GRADE BEAM CONCRETE**				
0960	Grade beam				
0980	2500# or 3000# concrete				
1000	By chute	C.Y.	12.50	98.00	111
1040	By pump	"	47.50	98.00	146
1060	By hand buggy	"	37.50	98.00	136
1150	5000# concrete				
1160	By chute	C.Y.	12.50	110	123
1190	By pump	"	47.50	110	158
1200	By hand buggy	"	37.50	110	148
03380.53	**PILE CAP CONCRETE**				
0970	Pile cap				
0980	2500# or 3000 concrete				
1000	By chute	C.Y.	12.50	00.00	111
1010	By pump	"	54.00	98.00	152
1020	By hand buggy	"	37.50	98.00	136
3980	5000# concrete				
4010	By chute	C.Y.	12.50	110	123
4020	By pump	"	54.00	110	164
4030	By hand buggy	"	37.50	110	148

		UNIT	LABOR	MAT.	TOTAL
03380.55	**SLAB/MAT CONCRETE**				
0960	Slab on grade				
0980	2500# or 3000# concrete				
1000	By chute	C.Y.	9.38	98.00	107
1020	By pump	"	27.25	98.00	125
1030	By hand buggy	"	25.00	98.00	123
3980	5000# concrete				
4010	By chute	C.Y.	9.38	110	119
4030	By pump	"	27.25	110	137
4040	By hand buggy	"	25.00	110	135
03380.58	**SIDEWALKS**				
6000	Walks, cast in place with wire mesh, base not incl.				
6010	4" thick	S.F.	1.25	1.34	2.59
6020	5" thick	"	1.50	1.80	3.30
6030	6" thick	"	1.87	2.26	4.13
03380.60	**STAIR CONCRETE**				
0960	Stairs				
0980	2500# or 3000# concrete				
1000	By chute	C.Y.	12.50	98.00	111
1030	By pump	"	54.00	98.00	152
1040	By hand buggy	"	37.50	98.00	136
2100	3500# or 4000# concrete				
2120	By chute	C.Y.	12.50	100	113
2160	By pump	"	54.00	100	154
2180	By hand buggy	"	37.50	100	138
4000	5000# concrete				
4010	By chute	C.Y.	12.50	110	123
4030	By pump	"	54.00	110	164
4040	By hand buggy	"	37.50	110	148
03380.65	**WALL CONCRETE**				
0940	Walls				
0960	2500# or 3000# concrete				
0980	To 4'				
1000	By chute	C.Y.	10.75	98.00	109
1010	By pump	"	58.00	98.00	156
1020	To 8'				
1040	By pump	C.Y.	63.00	98.00	161
2960	3500# or 4000# concrete				
2980	To 4'				
3000	By chute	C.Y.	10.75	100	111
3030	By pump	"	58.00	100	158
3060	To 8'				
3100	By pump	C.Y.	63.00	100	163
8480	Filled block (CMU)				
8490	3000# concrete, by pump				
8500	4" wide	S.F.	2.71	0.36	3.07
8510	6" wide	"	3.16	0.85	4.01
8520	8" wide	"	3.80	1.31	5.11
8530	10" wide	"	4.47	1.75	6.22
8540	12" wide	"	5.43	2.23	7.66
8560	Pilasters, 3000# concrete	C.F.	76.00	5.14	81.14
8700	Wall cavity, 2" thick, 3000# concrete	S.F.	2.53	0.94	3.47

DIVISION # 03 CONCRETE

		UNIT	LABOR	MAT.	TOTAL
03550.10	**CONCRETE TOPPINGS**				
1000	Gypsum fill				
1020	2" thick	S.F.	0.39	1.61	2.00
1040	2-1/2" thick	"	0.40	1.84	2.24
1060	3" thick	"	0.41	2.25	2.66
1080	3-1/2" thick	"	0.42	2.58	3.00
1100	4" thick	"	0.47	3.00	3.47
2000	Formboard				
2020	Mineral fiber board				
2040	1" thick	S.F.	0.93	1.44	2.37
2060	1-1/2" thick	"	1.07	3.78	4.85
2070	Cement fiber board				
2080	1" thick	S.F.	1.25	1.12	2.37
2100	1-1/2" thick	"	1.44	1.44	2.88
2110	Glass fiber board				
2120	1" thick	S.F.	0.93	1.77	2.70
2140	1-1/2" thick	"	1.07	2.39	3.46
4000	Poured deck				
4010	Vermiculite or perlite				
4020	1 to 4 mix	C.Y.	63.00	150	213
4040	1 to 6 mix	"	58.00	140	198
4050	Vermiculite or perlite				
4060	2" thick				
4080	1 to 4 mix	S.F.	0.40	1.44	1.84
4100	1 to 6 mix	"	0.36	1.05	1.41
4200	3" thick				
4220	1 to 4 mix	S.F.	0.58	1.96	2.54
4240	1 to 6 mix	"	0.54	1.55	2.09
6000	Concrete plank, lightweight				
6020	2" thick	S.F.	3.17	7.76	10.93
6040	2-1/2" thick	"	3.17	7.95	11.12
6080	3-1/2" thick	"	3.52	8.28	11.80
6100	4" thick	"	3.52	8.61	12.13
6500	Channel slab, lightweight, straight				
6520	2-3/4" thick	S.F.	3.17	6.26	9.43
6540	3-1/2" thick	"	3.17	6.53	9.70
6560	3-3/4" thick	"	3.17	6.93	10.10
6580	4-3/4" thick	"	3.52	8.79	12.31
7000	Gypsum plank				
7020	2" thick	S.F.	3.17	2.92	6.09
7040	3" thick	"	3.17	3.05	6.22
8000	Cement fiber, T and G planks				
8020	1" thick	S.F.	2.88	1.63	4.51
8040	1-1/2" thick	"	2.88	1.69	4.57
8060	2" thick	"	3.17	1.98	5.15
8080	2-1/2" thick	"	3.17	2.08	5.25
8100	3" thick	"	3.17	2.73	5.90
8120	3-1/2" thick	"	3.52	3.12	6.64
8140	4" thick	"	3.52	3.54	7.06
03730.10	**CONCRETE REPAIR**				
0090	Epoxy grout floor patch, 1/4" thick	S.F.	3.75	6.06	9.81
0100	Grout, epoxy, 2 component system	C.F.			290

DIVISION # 03 CONCRETE

		UNIT	LABOR	MAT.	TOTAL
03730.10	**CONCRETE REPAIR, Cont'd...**				
0110	Epoxy sand	BAG			19.75
0120	Epoxy modifier	GAL			130
0140	Epoxy gel grout	S.F.	37.50	2.95	40.45
0150	Injection valve, 1 way, threaded plastic	EA.	7.50	8.11	15.61
0155	Grout crack seal, 2 component	C.F.	37.50	680	718
0160	Grout, non shrink	"	37.50	70.00	108
0165	Concrete, epoxy modified				
0170	Sand mix	C.F.	15.00	110	125
0180	Gravel mix	"	14.00	85.00	99.00
0190	Concrete repair				
0195	Soffit repair				
0200	16" wide	L.F.	7.50	3.45	10.95
0210	18" wide	"	7.81	3.62	11.43
0220	24" wide	"	8.33	4.26	12.59
0230	30" wide	"	8.93	4.89	13.82
0240	32" wide	"	9.38	5.25	14.63
0245	Edge repair				
0250	2" spall	L.F.	9.38	1.62	11.00
0260	3" spall	"	9.87	1.62	11.49
0270	4" spall	"	10.25	1.72	11.97
0280	6" spall	"	10.50	1.81	12.31
0290	8" spall	"	11.00	1.89	12.89
0300	9" spall	"	12.50	1.95	14.45
0330	Crack repair, 1/8" crack	"	3.75	3.20	6.95
5000	Reinforcing steel repair				
5005	1 bar, 4 ft				
5010	#4 bar	L.F.	6.71	0.48	7.19
5012	#5 bar	"	6.71	0.67	7.38
5014	#6 bar	"	7.16	0.83	7.99
5016	#8 bar	"	7.16	1.49	8.65
5020	#9 bar	"	7.67	1.90	9.57
5030	#11 bar	"	7.67	2.96	10.63
7010	Form fabric, nylon				
7020	18" diameter	L.F.			12.75
7030	20" diameter	"			13.00
7040	24" diameter	"			21.25
7050	30" diameter	"			21.75
7060	36" diameter	"			25.00
7100	Pile repairs				
7105	Polyethylene wrap				
7108	30 mil thick				
7110	60" wide	S.F.	12.50	13.25	25.75
7120	72" wide	"	15.00	14.25	29.25
7125	60 mil thick				
7130	60" wide	S.F.	12.50	15.75	28.25
7140	80" wide	"	17.00	18.25	35.25
8010	Pile spall, average repair 3'				
8020	18" x 18"	EA.	31.25	41.00	72.25
8030	20" x 20"	"	37.50	55.00	92.50

© 2008 By Design & Construction Resources

03 - 13

		UNIT	COST
03999.10	**CONCRETE**		
1000	FOOTINGS (Incl. exc. with steel)		
1010	By L.F - Continuous 24" x 12" (3 # 4 Rods)	L.F.	24.75
1020	36" x 12" (4 # 4 Rods)	"	31.25
1030	20" x 10" (2 # 5 Rods)	"	21.00
1040	16" x 8" (2 # 4 Rods)	"	17.85
1050	Pad 24" x 24" x 12" (4 # 5 E.W.)	EA.	86.50
1060	36" x 36" x 14" (6 # 5 E.W.)	"	177
1070	48" x 48" x 16" (8 # 5 E.W.)	"	310
1100	By C.Y. - Continuous 24" x 12" (3 # 4 Rods)	C.Y.	310
1110	36" x 12" (4 # 4 Rods)	"	320
1120	20" x 10" (2 # 5 Rods)	"	340
1130	16" x 8" (2 # 4 Rods)	"	410
1140	Pad 24" x 24" x 12" (4 # 5 E.W.)	"	520
1150	36" x 36" x 14" (6 # 5 E.W.)	"	480
1160	48" x 48" x 16" (8 # 5 E.W.)	"	470
1200	WALLS (#5 Rods 12" O.C. - 1 Face)		
1210	By S.F. - 8" Wall (# 5 12" O.C. E.W.)	S.F.	17.79
1220	12" Wall (# 5 12" O.C. E.W.)	"	19.15
1230	16" Wall (# 5 12" O.C. E.W.)	"	19.15
1300	Add for Steel - 2 Faces - 12" Wall	"	2.26
1310	Add for Pilastered Wall - 24" O.C.	"	1.08
1320	Add for Retaining or Battered Type	"	3.14
1330	Add for Curved Walls	"	5.86
1340	By C.Y. - 8" Wall (# 5 12" O.C. E.W.)	C.Y.	570
1350	12" Wall (# 5 12" O.C. E.W.)	"	470
1360	16" Wall (# 5 12" O.C. E.W.)	"	430
1400	Add for Steel - 2 Faces - 12" Wall	"	63.75
1410	Add for Pilastered Wall - 24" O.C.	"	28.00
1420	Add for Retaining or Battered Type	"	87.00
1430	Add for Curved Walls	"	158
1500	COLUMNS		
1510	By L.F. - Square Cornered 8" x 8" (4 # 8 Rods)	L.F.	41.00
1520	12" x 12" (6 # 8 Rods)	"	69.25
1530	16" x 16" (6 # 10 Rods)	"	90.25
1540	20" x 20" (8 # 20 Rods)	"	129
1550	24" x 24" (10 # 11 Rods)	"	154
1600	Round 8" (4 # 8 Rods)	"	29.75
1610	12" (6 # 8 Rods)	"	47.00
1620	16" (6 # 10 Rods)	"	63.50
1630	20" (8 # 20 Rods)	"	89.75
1640	24" (10 # 11 Rods)	"	130
1700	By C.Y. - Sq Cornered 8" x 8" (4 # 8 Rods)	C.Y.	2,370
1710	12" x 12" (6 # 8 Rods)	"	2,190
1720	16" x 16" (6 # 10 Rods)	"	1,550
1730	20" x 20" (8 # 20 Rods)	"	1,380
1740	24" x 24" (10 # 11 Rods)	"	990
1800	Round 8" (4 # 8 Rods)	"	2,190
1810	12" (6 # 8 Rods)	"	1,440
1820	16" (6 # 10 Rods)	"	1,330
1830	20" (8 # 20 Rods)	"	1,210
1840	24" (10 # 11 Rods)	"	1,160
1900	BEAMS		

		UNIT	COST
03999.10	**CONCRETE, Cont'd...**		
1910	By L.F. - Spandrel 12" x 48" (33 # rebar)	L.F.	156
1920	12" x 42" (26 # Rein. Steel)	"	144
1930	12" x 36" (21 # Rein. Steel)	"	113
1940	12" x 30" (15 # Rein. Steel)	"	107
1950	8" x 48" (26 # Rein. Steel)	"	140
1960	8" x 42" (21 # Rein. Steel)	"	127
1970	8" x 36" (16 # Rein. Steel)	"	115
2000	Interior 16" x 30" (24 # rebar)	"	91.25
2010	16" x 24" (20 # Rein. Steel)	"	90.50
2020	12" x 30" (17 # Rein. Steel)	"	106
2030	12" x 24" (14 # Rein. Steel)	"	75.50
2040	12" x 16" (12 # Rein. Steel)	"	64.25
2050	8" x 24" (13 # Rein. Steel)	"	75.50
2060	8" x 16" (10 # Rein. Steel)	"	56.50
2100	By C.Y. - Spandrel 12" x 48" (33 # rebar)	C.Y.	930
2110	12" x 42" (26 # Rein. Steel)	"	990
2120	12" x 36" (21 # Rein. Steel)	"	1,040
2130	12" x 30" (15 # Rein. Steel)	"	1,100
2140	8" x 48" (26 # Rein. Steel)	"	1,100
2150	8" x 42" (21 # Rein. Steel)	"	1,200
2160	8" x 36" (16 # Rein. Steel)	"	1,250
2200	Interior 16" x 30" (24 # rebar)	"	610
2210	16" x 24" (20 # Rein. Steel)	"	900
2220	12" x 30" (17 # Rein. Steel)	"	990
2230	12" x 24" (14 # Rein. Steel)	"	960
2240	12" x 16" (12 # Rein. Steel)	"	1,090
2250	8" x 24" (13 # Rein. Steel)	"	1,210
2260	8" x 16" (10 # Rein. Steel)	"	1,270
2300	SLABS (With Reinf. Steel)		
2310	By S.F. - Solid		
2320	4" Thick	S.F.	11.05
2330	5" Thick	"	12.95
2340	6" Thick	"	14.17
2350	7" Thick	"	15.81
2360	8" Thick	"	15.26
2400	Deduct for Post Tensioned Slabs	"	1.01
2410	Pan	"	
2420	Joist -20" Pan - 10" x 2"	"	15.35
2430	12" x 2"	"	17.34
2440	30" Pan - 10" x 2 1/2"	"	15.51
2450	12" x 2 1/2"	"	17.34
2500	Dome -19" x 19" - 10" x 2"	"	16.95
2510	12" x 2"	"	17.57
2520	30" x 30" - 10" x 2 1/2"	"	14.97
2530	12" x 2 1/2"	"	16.63
2600	By C.Y. - Solid		
2610	4" Thick	C.Y.	870
2620	5" Thick	"	830
2630	6" Thick	"	810
2640	8" Thick	"	780
2700	COMBINED COLUMNS, BEAMS AND SLABS		
2710	20' Span	S.F.	23.00

		UNIT	COST
03999.10	**CONCRETE, Cont'd...**		
2720	30' Span	S.F.	24.50
2730	40' Span	"	26.00
2740	50' Span	"	27.75
2750	60' Span	"	47.25
2800	STAIRS (Including Landing)		
2810	By EA. - 4' Wide 10' Floor Heights 16 Risers	EA.	210
2820	5' Wide 10' Floor Heights 16 Risers	"	240
2830	6' Wide 10' Floor Heights 16 Risers	"	280
2840	By C.Y. - 4' Wide 10' Floor Heights 16 Risers	C.Y.	1,290
2850	5' Wide 10' Floor Heights 16 Risers	"	1,330
2860	6' Wide 10' Floor Heights 16 Risers	"	1,300
2900	SLABS ON GROUND		
2910	4" Concrete Slab		
2920	(6 6/10 - 10 Mesh - 5 1/2 Sack Concrete,		
2930	Trowel Finished, Cured & Truck Chuted)	S.F.	4.01
2940	Add per Inch of Concrete	"	0.67
2950	Add per Sack of Cement	"	0.17
2960	Deduct for Float Finish	"	0.08
2970	Deduct for Brush or Broom Finish	"	0.06
3000	Add for Runway and Buggied Concrete	"	0.21
3010	Add for Vapor Barrier (4 mil)	"	0.17
3020	Add for Sub - Floor Fill (4" sand/gravel)	"	0.50
3030	Add for Change to 6 6/8 - 8 Mesh	"	0.09
3040	Add for Change to 6 6/6 - 6 Mesh	"	0.17
3050	Add for Sloped Slab	"	0.24
3060	Add for Edge Strip (sidewalk area)	"	0.27
3100	Add for ½" Expansion Joint (20' O.C.)	"	0.08
3110	Add for Control Joints (keyed/ dep.)	"	0.19
3120	Add for Control Joints (joint filled)	"	0.21
3130	Add for Control Joints - Saw Cut (20' O.C.)	"	0.36
3140	Add for Floor Hardener (1 coat)	"	0.17
3150	Add for Exposed Aggregate - Washed Added	"	0.50
3200	Retarding Added	"	0.46
3210	Seeding Added	"	0.49
3220	Add for Light Weight Aggregates	"	0.50
3230	Add for Heavy Weight Aggregates	"	0.34
3240	Add for Winter Production Loss/Cost	"	0.42
3300	TOPPING SLABS		
3310	2" Concrete	S.F.	3.31
3320	3" Concrete	"	4.17
3330	No Mesh or Hoisting		
3400	PADS & PLATFORMS (Including Form Work & Reinforcing)		
3410	4"	S.F.	7.90
3420	6"	"	10.01
3500	PRECAST CONCRETE ITEMS		
3510	Curbs 6" x 10" x 8"	L.F.	13.65
3520	Sills & Stools 6"	"	30.75
3530	Splash Blocks 3" x 16"	EA.	74.50
3600	MISCELLANEOUS ADDITIONS TO ABOVE CONCRETE		
3610	Abrasives - Carborundum - Grits	S.F.	1.01
3620	Strips	L.F.	3.36
3630	Bushhammer - Green Concrete	S.F.	2.17

		UNIT	COST
03999.10	**CONCRETE, Cont'd...**		
3640	Cured Concrete	S.F.	2.97
3650	Chamfers - Plastic 3/4"	L.F.	1.19
3660	Wood 3/4"	"	0.67
3670	Metal 3/4"	"	1.31
3680	Colors Dust On	S.F.	0.86
3690	Integral (Top 1")	"	1.62
3700	Control Joints - Asphalt 1/2" x 4"	L.F.	0.96
3710	1/2" x 6"	"	1.11
3720	PolyFoam 1/2" x 4"	"	1.19
3730	Dovetail Slots 22 Ga	"	1.73
3740	24 Ga	"	1.13
3800	Hardeners Acrylic and Urethane	S.F.	0.29
3810	Epoxy	"	0.38
3820	Joint Sealers Epoxy	L.F.	3.37
3830	Rubber Asphalt	"	1.76
3840	Moisture Proofing - Polyethylene 4 mil	S.F.	0.19
3850	6 mil	"	0.24
3900	Non-Shrink Grouts - Non - Metallic	C.F.	46.50
3910	Aluminum Oxide	"	33.75
3920	Iron Oxide	"	50.00
4000	Reglets Flashing	L.F.	1.87
4010	Sand Blast - Light	S.F.	1.61
4020	Heavy	"	2.95
4030	Shelf Angle Inserts 5/8"	L.F.	6.05
4040	Stair Nosings Steel - Galvanized	"	8.15
4050	Tongue & Groove Joint Forms - Asphalt 5 1/2"	"	2.48
4060	Wood 5 1/2"	"	2.02
4070	Metal 5 1/2"	"	2.59
4100	Treads - Extruded Aluminum	"	11.19
4110	Cast Iron	"	12.64
4120	Water Stops Center Bulb - Rubber - 6"	"	12.81
4130	9"	"	23.75
4140	Polyethylene - 6"	"	6.23
4150	9"	"	6.32

Design & Construction Resources

TABLE OF CONTENTS   PAGE

04100.10 - MASONRY GROUT   04-2
04150.10 - MASONRY ACCESSORIES   04-2
04150.20 - MASONRY CONTROL JOINTS   04-4
04150.50 - MASONRY FLASHING   04-4
04210.10 - BRICK MASONRY   04-4
04210.20 - STRUCTURAL TILE   04-5
04210.60 - PAVERS, MASONRY   04-6
04220.10 - CONCRETE MASONRY UNITS   04-6
04220.90 - BOND BEAMS & LINTELS   04-8
04240.10 - CLAY TILE   04-9
04270.10 - GLASS BLOCK   04-10
04295.10 - PARGING/MASONRY PLASTER   04-10
04400.10 - STONE   04-10
QUICK ESTIMATING   04A-13

		UNIT	LABOR	MAT.	TOTAL
04100.10	**MASONRY GROUT**				
0100	Grout, non shrink, non-metallic, trowelable	C.F.	1.40	4.95	6.35
2110	Grout door frame, hollow metal				
2120	Single	EA.	53.00	12.25	65.25
2140	Double	"	56.00	17.25	73.25
2980	Grout-filled concrete block (CMU)				
3000	4" wide	S.F.	1.76	0.33	2.09
3020	6" wide	"	1.92	0.86	2.78
3040	8" wide	"	2.11	1.27	3.38
3060	12" wide	"	2.22	2.09	4.31
3070	Grout-filled individual CMU cells				
3090	4" wide	L.F.	1.05	0.27	1.32
3100	6" wide	"	1.05	0.37	1.42
3120	8" wide	"	1.05	0.49	1.54
3140	10" wide	"	1.20	0.61	1.81
3160	12" wide	"	1.20	0.74	1.94
4000	Bond beams or lintels, 8" deep				
4010	6" thick	L.F.	1.72	0.74	2.46
4020	8" thick	"	1.90	0.99	2.89
4040	10" thick	"	2.11	1.24	3.35
4060	12" thick	"	2.37	1.48	3.85
5000	Cavity walls				
5020	2" thick	S.F.	2.53	0.82	3.35
5040	3" thick	"	2.53	1.24	3.77
5060	4" thick	"	2.71	1.65	4.36
5080	6" thick	"	3.16	2.47	5.63
04150.10	**MASONRY ACCESSORIES**				
0200	Foundation vents	EA.	18.75	31.00	49.75
1010	Bar reinforcing				
1015	Horizontal				
1020	#3 - #4	Lb.	1.86	0.56	2.42
1030	#5 - #6	"	1.55	0.56	2.11
1035	Vertical				
1040	#3 - #4	Lb.	2.33	0.56	2.89
1050	#5 - #6	"	1.86	0.56	2.42
1100	Horizontal joint reinforcing				
1105	Truss type				
1110	4" wide, 6" wall	L.F.	0.18	0.20	0.38
1120	6" wide, 8" wall	"	0.19	0.20	0.39
1130	8" wide, 10" wall	"	0.20	0.25	0.45
1140	10" wide, 12" wall	"	0.21	0.25	0.46
1150	12" wide, 14" wall	"	0.22	0.30	0.52
1155	Ladder type				
1160	4" wide, 6" wall	L.F.	0.18	0.14	0.32
1170	6" wide, 8" wall	"	0.19	0.16	0.35
1180	8" wide, 10" wall	"	0.20	0.17	0.37
1190	10" wide, 12" wall	"	0.20	0.20	0.40
2000	Rectangular wall ties				
2005	3/16" dia., galvanized				
2010	2" x 6"	EA.	0.77	0.22	0.99
2020	2" x 8"	"	0.77	0.23	1.00
2040	2" x 10"	"	0.77	0.26	1.03

		UNIT	LABOR	MAT.	TOTAL
04150.10	**MASONRY ACCESSORIES, Cont'd...**				
2050	2" x 12"	EA.	0.77	0.30	1.07
2060	4" x 6"	"	0.93	0.25	1.18
2070	4" x 8"	"	0.93	0.27	1.20
2080	4" x 10"	"	0.93	0.36	1.29
2090	4" x 12"	"	0.93	0.42	1.35
2095	1/4" dia., galvanized				
2100	2" x 6"	EA.	0.77	0.41	1.18
2110	2" x 8"	"	0.77	0.45	1.22
2120	2" x 10"	"	0.77	0.51	1.28
2130	2" x 12"	"	0.77	0.58	1.35
2140	4" x 6"	"	0.93	0.47	1.40
2150	4" x 8"	"	0.93	0.51	1.44
2160	4" x 10"	"	0.93	0.58	1.51
2170	4" x 12"	"	0.93	0.61	1.54
2200	"Z" type wall ties, galvanized				
2215	6" long				
2220	1/8" dia.	EA.	0.77	0.22	0.99
2230	3/16" dia.	"	0.77	0.23	1.00
2240	1/4" dia.	"	0.77	0.25	1.02
2245	8" long				
2250	1/8" dia.	EA.	0.77	0.23	1.00
2260	3/16" dia.	"	0.77	0.25	1.02
2270	1/4" dia.	"	0.77	0.26	1.03
2275	10" long				
2280	1/8" dia.	EA.	0.77	0.25	1.02
2290	3/16" dia.	"	0.77	0.27	1.04
2300	1/4" dia.	"	0.77	0.32	1.09
3000	Dovetail anchor slots				
3015	Galvanized steel, filled				
3020	24 ga.	L.F.	1.16	0.58	1.74
3040	20 ga.	"	1.16	0.73	1.89
3060	16 oz. copper, foam filled	"	1.16	1.45	2.61
3100	Dovetail anchors				
3115	16 ga.				
3120	3-1/2" long	EA.	0.77	0.17	0.94
3140	5-1/2" long	"	0.77	0.20	0.97
3150	12 ga.				
3160	3-1/2" long	EA.	0.77	0.22	0.99
3180	5-1/2" long	"	0.77	0.44	1.21
3200	Dovetail, triangular galvanized ties, 12 ga.				
3220	3" x 3"	EA.	0.77	0.40	1.17
3240	5" x 5"	"	0.77	0.43	1.20
3260	7" x 7"	"	0.77	0.49	1.26
3280	7" x 9"	"	0.77	0.51	1.28
3400	Brick anchors				
3420	Corrugated, 3-1/2" long				
3440	16 ga.	EA.	0.77	0.15	0.92
3460	12 ga.	"	0.77	0.27	1.04
3500	Non-corrugated, 3-1/2" long				
3520	16 ga.	EA.	0.77	0.21	0.98
3540	12 ga.	"	0.77	0.38	1.15
3580	Cavity wall anchors, corrugated, galvanized				

		UNIT	LABOR	MAT.	TOTAL
04150.10	**MASONRY ACCESSORIES, Cont'd...**				
3600	5" long				
3620	16 ga.	EA.	0.77	0.49	1.26
3640	12 ga.	"	0.77	0.72	1.49
3660	7" long				
3680	28 ga.	EA.	0.77	0.53	1.30
3700	24 ga.	"	0.77	0.67	1.44
3720	22 ga.	"	0.77	0.69	1.46
3740	16 ga.	"	0.77	0.77	1.54
3800	Mesh ties, 16 ga., 3" wide				
3820	8" long	EA.	0.77	0.64	1.41
3840	12" long	"	0.77	0.72	1.49
3860	20" long	"	0.77	0.99	1.76
3900	24" long	"	0.77	1.09	1.86
04150.20	**MASONRY CONTROL JOINTS**				
1000	Control joint, cross shaped PVC	L.F.	1.16	2.17	3.33
1010	Closed cell joint filler				
1020	1/2"	L.F.	1.16	0.38	1.54
1040	3/4"	"	1.16	0.77	1.93
1070	Rubber, for				
1080	4" wall	L.F.	1.16	2.50	3.66
1090	6" wall	"	1.22	3.10	4.32
1100	8" wall	"	1.29	3.74	5.03
1110	PVC, for				
1120	4" wall	L.F.	1.16	1.30	2.46
1140	6" wall	"	1.22	2.20	3.42
1160	8" wall	"	1.29	3.33	4.62
04150.50	**MASONRY FLASHING**				
0080	Through-wall flashing				
1000	5 oz. coated copper	S.F.	3.89	3.35	7.24
1020	0.030" elastomeric	"	3.11	1.10	4.21
04210.10	**BRICK MASONRY**				
0100	Standard size brick, running bond				
1000	Face brick, red (6.4/sf)				
1020	Veneer	S.F.	7.78	5.39	13.17
1030	Cavity wall	"	6.67	5.39	12.06
1040	9" solid wall	"	13.25	10.75	24.00
1200	Common brick (6.4/sf)				
1210	Select common for veneers	S.F.	7.78	3.74	11.52
1215	Back-up				
1220	4" thick	S.F.	5.84	3.41	9.25
1230	8" thick	"	9.34	6.82	16.16
1235	Firewall				
1240	12" thick	S.F.	15.50	10.25	25.75
1250	16" thick	"	21.25	13.75	35.00
1300	Glazed brick (7.4/sf)				
1310	Veneer	S.F.	8.49	12.00	20.49
1400	Buff or gray face brick (6.4/sf)				
1410	Veneer	S.F.	7.78	5.98	13.76
1420	Cavity wall	"	6.67	5.98	12.65
1500	Jumbo or oversize brick (3/sf)				
1510	4" veneer	S.F.	4.67	4.12	8.79

		UNIT	LABOR	MAT.	TOTAL
04210.10	**BRICK MASONRY, Cont'd...**				
1530	4" back-up	S.F.	3.89	4.12	8.01
1540	8" back-up	"	6.67	8.25	14.92
1550	12" firewall	"	11.75	12.25	24.00
1560	16" firewall	"	15.50	16.50	32.00
1600	Norman brick, red face, (4.5/sf)				
1620	4" veneer	S.F.	5.84	5.94	11.78
1640	Cavity wall	"	5.19	5.94	11.13
3000	Chimney, standard brick, including flue				
3020	16" x 16"	L.F.	46.75	26.50	73.25
3040	16" x 20"	"	46.75	29.50	76.25
3060	16" x 24"	"	46.75	38.00	84.75
3080	20" x 20"	"	58.00	38.25	96.25
3100	20" x 24"	"	58.00	40.75	98.75
3120	20" x 32"	"	67.00	49.75	117
4000	Window sill, face brick on edge	"	11.75	2.97	14.72
04210.20	**STRUCTURAL TILE**				
5000	Structural glazed tile				
5010	6T series, 5-1/2" x 12"				
5020	Glazed on one side				
5040	2" thick	S.F.	4.67	7.26	11.93
5060	4" thick	"	4.67	8.60	13.27
5080	6" thick	"	5.19	13.50	18.69
5100	8" thick	"	5.84	16.50	22.34
5200	Glazed on two sides				
5220	4" thick	S.F.	5.84	15.00	20.84
5240	6" thick	"	6.67	17.25	23.92
5500	Special shapes				
5510	Group 1	S.F.	9.34	11.25	20.59
5520	Group 2	"	9.34	14.25	23.59
5530	Group 3	"	9.34	15.75	25.09
5540	Group 4	"	9.34	27.50	36.84
5550	Group 5	"	9.34	33.00	42.34
5600	Fire rated				
5620	4" thick, 1 hr rating	S.F.	4.67	21.00	25.67
5640	6" thick, 2 hr rating	"	5.19	25.50	30.69
6000	8W series, 8" x 16"				
6010	Glazed on one side				
6020	2" thick	S.F.	3.11	8.10	11.21
6040	4" thick	"	3.11	8.80	11.91
6060	6" thick	"	3.59	12.25	15.84
6080	8" thick	"	3.59	14.25	17.84
6100	Glazed on two sides				
6120	4" thick	S.F.	3.89	15.00	18.89
6140	6" thick	"	4.67	18.00	22.67
6160	8" thick	"	4.67	25.50	30.17
6200	Special shapes				
6220	Group 1	S.F.	6.67	13.50	20.17
6230	Group 2	"	6.67	16.50	23.17
6240	Group 3	"	6.67	18.00	24.67
6250	Group 4	"	6.67	30.00	36.67
6260	Group 5	"	6.67	75.00	81.67

		UNIT	LABOR	MAT.	TOTAL
04210.20	**STRUCTURAL TILE, Cont'd...**				
6270	Fire rated				
6290	4" thick, 1 hr rating	S.F.	6.67	22.50	29.17
6300	6" thick, 2 hr rating	"	6.67	27.00	33.67
04210.60	**PAVERS, MASONRY**				
4000	Brick walk laid on sand, sand joints				
4020	Laid flat, (4.5 per sf)	S.F.	5.19	3.78	8.97
4040	Laid on edge, (7.2 per sf)	"	7.78	6.05	13.83
5000	Precast concrete patio blocks				
5005	2" thick				
5010	Natural	S.F.	1.55	2.68	4.23
5020	Colors	"	1.55	3.74	5.29
5080	Exposed aggregates, local aggregate				
5100	Natural	S.F.	1.55	3.13	4.68
5120	Colors	"	1.55	4.19	5.74
5130	Granite or limestone aggregate	"	1.55	5.74	7.29
5140	White tumblestone aggregate	"	1.55	4.47	6.02
6000	Stone pavers, set in mortar				
6005	Bluestone				
6008	1" thick				
6010	Irregular	S.F.	11.75	3.48	15.23
6020	Snapped rectangular	"	9.34	5.30	14.64
6060	1-1/2" thick, random rectangular	"	11.75	6.08	17.83
6070	2" thick, random rectangular	"	13.25	7.19	20.44
6090	Slate				
6100	Natural cleft				
6110	Irregular, 3/4" thick	S.F.	13.25	3.48	16.73
6115	Random rectangular				
6120	1-1/4" thick	S.F.	11.75	7.54	19.29
6130	1-1/2" thick	"	13.00	8.51	21.51
7000	Granite blocks				
7010	3" thick, 3" to 6" wide				
7020	4" to 12" long	S.F.	15.50	17.00	32.50
7030	6" to 15" long	"	13.25	17.00	30.25
9000	Flagstone pavers				
9010	Random sizes, 1-4 s.f., tumbled cobble	S.F.	9.34	15.50	24.84
9020	Tumbled patio	"	9.34	16.50	25.84
9040	Saw-cut Flagstone Tiles				
9060	12"x12", assorted colors	S.F.	6.67	11.00	17.67
9080	18"x18", assorted colors	"	5.84	11.00	16.84
9100	24"x24", assorted colors	"	5.19	11.00	16.19
9800	Crushed stone, white marble, 3" thick	"	0.75	1.60	2.35
04220.10	**CONCRETE MASONRY UNITS**				
0110	Hollow, load bearing				
0120	4"	S.F.	3.46	1.60	5.06
0140	6"	"	3.59	2.01	5.60
0160	8"	"	3.89	2.59	6.48
0180	10"	"	4.24	3.47	7.71
0190	12"	"	4.67	3.66	8.33
0280	Solid, load bearing				
0300	4"	S.F.	3.46	2.15	5.61
0320	6"	"	3.59	2.57	6.16

		UNIT	LABOR	MAT.	TOTAL
04220.10	**CONCRETE MASONRY UNITS, Cont'd...**				
0340	8"	S.F.	3.89	2.98	6.87
0360	10"	"	4.24	4.04	8.28
0380	12"	"	4.67	4.49	9.16
0480	Back-up block, 8" x 16"				
0500	2"	S.F.	2.66	1.32	3.98
0540	4"	"	2.74	1.60	4.34
0560	6"	"	2.92	2.01	4.93
0580	8"	"	3.11	2.59	5.70
0600	10"	"	3.33	3.47	6.80
0620	12"	"	3.59	3.66	7.25
0980	Foundation wall, 8" x 16"				
1000	6"	S.F.	3.33	2.01	5.34
1030	8"	"	3.59	2.59	6.18
1040	10"	"	3.89	3.47	7.36
1050	12"	"	4.24	3.66	7.90
1055	Solid				
1060	6"	S.F.	3.59	2.57	6.16
1070	8"	"	3.89	2.98	6.87
1080	10"	"	4.24	4.04	8.28
1100	12"	"	4.67	4.49	9.16
1480	Exterior, styrofoam inserts, standard weight, 8" x 16"				
1500	6"	S.F.	3.59	2.36	5.95
1530	8"	"	3.89	2.47	6.36
1540	10"	"	4.24	2.91	7.15
1550	12"	"	4.67	3.04	7.71
1580	Lightweight				
1600	6"	S.F.	3.59	3.13	6.72
1660	8"	"	3.89	3.80	7.69
1680	10"	"	4.24	4.80	9.04
1700	12"	"	4.67	5.08	9.75
1980	Acoustical slotted block				
2000	4"	S.F.	4.24	3.39	7.63
2020	6"	"	4.24	4.14	8.38
2040	8"	"	4.67	6.21	10.88
2050	Filled cavities				
2060	4"	S.F.	5.19	4.29	9.48
2070	6"	"	5.49	5.31	10.80
2080	8"	"	5.84	7.09	12.93
4000	Hollow, split face				
4020	4"	S.F.	3.46	3.25	6.71
4030	6"	"	3.59	3.77	7.36
4040	8"	"	3.89	3.96	7.85
4080	10"	"	4.24	4.43	8.67
4100	12"	"	4.67	4.73	9.40
4480	Split rib profile				
4500	4"	S.F.	4.24	2.83	7.07
4520	6"	"	4.24	3.27	7.51
4540	8"	"	4.67	3.56	8.23
4560	10"	"	4.67	3.90	8.57
4580	12"	"	4.67	4.25	8.92
4980	High strength block, 3500 psi				
5000	2"	S.F.	3.46	1.54	5.00

		UNIT	LABOR	MAT.	TOTAL
04220.10	**CONCRETE MASONRY UNITS, Cont'd...**				
5020	4"	S.F.	3.59	1.92	5.51
5030	6"	"	3.59	2.29	5.88
5040	8"	"	3.89	2.59	6.48
5050	10"	"	4.24	3.02	7.26
5060	12"	"	4.67	3.58	8.25
5500	Solar screen concrete block				
5505	4" thick				
5510	6" x 6"	S.F.	10.50	4.58	15.08
5520	8" x 8"	"	9.34	5.47	14.81
5530	12" x 12"	"	7.18	5.61	12.79
5540	8" thick				
5550	8" x 16"	S.F.	6.67	5.61	12.28
7000	Glazed block				
7020	Cove base, glazed 1 side, 2"	L.F.	5.19	10.50	15.69
7030	4"	"	5.19	10.75	15.94
7040	6"	"	5.84	11.00	16.84
7050	8"	"	5.84	11.75	17.59
7055	Single face				
7060	2"	S.F.	3.89	8.14	12.03
7080	4"	"	3.89	8.81	12.70
7090	6"	"	4.24	8.93	13.17
7100	8"	"	4.67	9.25	13.92
7105	10"	"	5.19	9.57	14.76
7110	12"	"	5.49	9.87	15.36
7115	Double face				
7120	4"	S.F.	4.91	13.00	17.91
7140	6"	"	5.19	14.00	19.19
7160	8"	"	5.84	14.50	20.34
7180	Corner or bullnose				
7200	2"	EA.	5.84	11.00	16.84
7240	4"	"	6.67	14.00	20.67
7260	6"	"	6.67	17.25	23.92
7280	8"	"	7.78	18.75	26.53
7290	10"	"	8.49	20.25	28.74
7300	12"	"	9.34	22.00	31.34
9500	Gypsum unit masonry				
9510	Partition blocks (12"x30")				
9515	Solid				
9520	2"	S.F.	1.86	1.01	2.87
9525	Hollow				
9530	3"	S.F.	1.86	1.02	2.88
9540	4"	"	1.94	1.16	3.10
9550	6"	"	2.12	1.25	3.37
9900	Vertical reinforcing				
9920	4' o.c., add 5% to labor				
9940	2'8" o.c., add 15% to labor				
9960	Interior partitions, add 10% to labor				
04220.90	**BOND BEAMS & LINTELS**				
0980	Bond beam, no grout or reinforcement				
0990	8" x 16" x				
1000	4" thick	L.F.	3.59	1.12	4.71

		UNIT	LABOR	MAT.	TOTAL
04220.90	**BOND BEAMS & LINTELS, Cont'd...**				
1040	6" thick	L.F.	3.73	1.70	5.43
1060	8" thick	"	3.89	1.95	5.84
1080	10" thick	"	4.06	2.42	6.48
1100	12" thick	"	4.24	2.75	6.99
6000	Beam lintel, no grout or reinforcement				
6010	8" x 16" x				
6020	10" thick	L.F.	4.67	5.58	10.25
6040	12" thick	"	5.19	7.09	12.28
6080	Precast masonry lintel				
7000	6 lf, 8" high x				
7020	4" thick	L.F.	7.78	5.36	13.14
7040	6" thick	"	7.78	6.85	14.63
7060	8" thick	"	8.49	7.75	16.24
7080	10" thick	"	8.49	9.25	17.74
7090	10 lf, 8" high x				
7100	4" thick	L.F.	4.67	6.74	11.41
7120	6" thick	"	4.67	8.31	12.98
7140	8" thick	"	5.19	9.25	14.44
7160	10" thick	"	5.19	12.50	17.69
8000	Steel angles and plates				
8010	Minimum	Lb.	0.66	0.66	1.32
8020	Maximum	"	1.16	0.88	2.04
8200	Various size angle lintels				
8205	1/4" stock				
8210	3" x 3"	L.F.	2.92	3.74	6.66
8220	3" x 3-1/2"	"	2.92	3.89	6.81
8225	3/8" stock				
8230	3" x 4"	L.F.	2.92	5.47	8.39
8240	3-1/2" x 4"	"	2.92	5.75	8.67
8250	4" x 4"	"	2.92	6.20	9.12
8260	5" x 3-1/2"	"	2.92	6.49	9.41
8262	6" x 3-1/2"	"	2.92	6.64	9.56
8265	1/2" stock				
8280	6" x 4"	L.F.	2.92	8.27	11.19
04240.10	**CLAY TILE**				
0100	Hollow clay tile, for back-up, 12" x 12"				
1000	Scored face				
1010	Load bearing				
1020	4" thick	S.F.	3.33	4.54	7.87
1040	6" thick	"	3.46	5.56	9.02
1060	8" thick	"	3.59	6.72	10.31
1080	10" thick	"	3.73	7.46	11.19
1100	12" thick	"	3.89	8.77	12.66
2000	Non-load bearing				
2020	3" thick	S.F.	3.22	4.25	7.47
2040	4" thick	"	3.33	4.38	7.71
2060	6" thick	"	3.46	4.54	8.00
2080	8" thick	"	3.59	5.85	9.44
2100	12" thick	"	3.89	6.44	10.33
4100	Partition, 12" x 12"				
4150	In walls				

		UNIT	LABOR	MAT.	TOTAL
04240.10	**CLAY TILE, Cont'd...**				
4201	3" thick	S.F.	3.89	3.67	7.56
4210	4" thick	"	3.89	4.25	8.14
4220	6" thick	"	4.06	4.68	8.74
4230	8" thick	"	4.24	6.14	10.38
4240	10" thick	"	4.44	7.32	11.76
4250	12" thick	"	4.67	7.76	12.43
4300	Clay tile floors				
4320	4" thick	S.F.	2.59	4.54	7.13
4330	6" thick	"	2.74	5.56	8.30
4340	8" thick	"	2.92	6.72	9.64
4350	10" thick	"	3.11	7.46	10.57
4360	12" thick	"	3.33	8.77	12.10
6000	Terra cotta				
6020	Coping, 10" or 12" wide, 3" thick	L.F.	9.34	14.75	24.09
04270.10	**GLASS BLOCK**				
1000	Glass block, 4" thick				
1040	6" x 6"	S.F.	15.50	23.50	39.00
1060	8" x 8"	"	11.75	15.00	26.75
1080	12" x 12"	"	9.34	18.75	28.09
8980	Replacement glass blocks, 4" x 8" x 8"				
9100	Minimum	S.F.	46.75	18.25	65.00
9120	Maximum	"	93.00	21.00	114
04295.10	**PARGING/MASONRY PLASTER**				
0080	Parging				
0100	1/2" thick	S.F.	3.11	0.27	3.38
0200	3/4" thick	"	3.89	0.34	4.23
0300	1" thick	"	4.67	0.45	5.12
04400.10	**STONE**				
0160	Rubble stone				
0180	Walls set in mortar				
0200	8" thick	S.F.	11.75	15.00	26.75
0220	12" thick	"	18.75	18.25	37.00
0420	18" thick	"	23.25	24.00	47.25
0440	24" thick	"	31.25	30.25	61.50
0445	Dry set wall				
0450	8" thick	S.F.	7.78	17.00	24.78
0455	12" thick	"	11.75	19.00	30.75
0460	18" thick	"	15.50	26.50	42.00
0465	24" thick	"	18.75	32.25	51.00
0480	Cut stone				
0490	Imported marble				
0510	Facing panels				
0520	3/4" thick	S.F.	18.75	39.00	57.75
0530	1-1/2" thick	"	21.25	59.00	80.25
0540	2-1/4" thick	"	26.00	68.00	94.00
0600	Base				
0610	1" thick				
0620	4" high	L.F.	23.25	17.75	41.00
0640	6" high	"	23.25	21.75	45.00
0700	Columns, solid				
0720	Plain faced	C.F.	310	120	430

04400.10 STONE, Cont'd...		UNIT	LABOR	MAT.	TOTAL
0740	Fluted	C.F.	310	320	630
0780	Flooring, travertine, minimum	S.F.	7.18	16.00	23.18
0800	Average	"	9.34	21.50	30.84
0820	Maximum	"	10.50	32.25	42.75
1000	Domestic marble				
1020	Facing panels				
1040	7/8" thick	S.F.	18.75	35.25	54.00
1060	1-1/2" thick	"	21.25	53.00	74.25
1080	2-1/4" thick	"	26.00	64.00	90.00
1500	Stairs				
1510	12" treads	L.F.	23.25	31.50	54.75
1520	6" risers	"	15.50	23.50	39.00
1525	Thresholds, 7/8" thick, 3' long, 4" to 6" wide				
1530	Plain	EA.	39.00	30.00	69.00
1540	Beveled	"	39.00	33.25	72.25
1545	Window sill				
1550	6" wide, 2" thick	L.F.	18.75	15.25	34.00
1555	Stools				
1560	5" wide, 7/8" thick	L.F.	18.75	22.25	41.00
1620	Limestone panels up to 12' x 5', smooth finish				
1630	2" thick	S.F.	8.45	23.50	31.95
1650	3" thick	"	8.45	27.50	35.95
1660	4" thick	"	8.45	39.00	47.45
1760	Miscellaneous limestone items				
1770	Steps, 14" wide, 6" deep	L.F.	31.25	75.00	106
1780	Coping, smooth finish	C.F.	15.50	100	116
1790	Sills, lintels, jambs, smooth finish	"	18.75	100	119
1800	Granite veneer facing panels, polished				
1810	7/8" thick				
1820	Black	S.F.	18.75	41.75	60.50
1840	Gray	"	18.75	36.25	55.00
1850	Base				
1860	4" high	L.F.	9.34	19.50	28.84
1870	6" high	"	10.50	23.50	34.00
1880	Curbing, straight, 6" x 16"	"	35.25	21.50	56.75
1890	Radius curbs, radius over 5'	"	47.00	26.25	73.25
1900	Ashlar veneer				
1905	4" thick, random	S.F.	18.75	35.75	54.50
1910	Pavers, 4" x 4" split				
1915	Gray	S.F.	9.34	35.00	44.34
1920	Pink	"	9.34	34.50	43.84
1930	Black	"	9.34	34.00	43.34
2000	Slate, panels				
2010	1" thick	S.F.	18.75	25.00	43.75
2020	2" thick	"	21.25	34.00	55.25
2030	Sills or stools				
2040	1" thick				
2060	6" wide	L.F.	18.75	11.75	30.50
2080	10" wide	"	20.25	19.00	39.25
2100	2" thick				
2120	6" wide	L.F.	21.25	19.00	40.25
2140	10" wide	"	23.25	31.50	54.75

		UNIT	**LABOR**	**MAT.**	**TOTAL**
04400.10	**STONE, Cont'd...**				
2200	Simulated masonry				
2210	Cultured stone veneer, 1-3/4" thick				
2220	Ledgestone, large-pattern	S.F.	4.67	5.06	9.73
2230	Small-pattern	"	6.22	5.06	11.28
2240	Drystack	"	5.30	5.06	10.36
2250	Castlestone	"	4.76	5.74	10.50
2260	Fieldstone	"	5.19	5.74	10.93
2270	Limestone	"	4.24	5.74	9.98
2280	Split-face	"	4.67	5.74	10.41
2290	Stream stone (river rock)	"	4.91	5.74	10.65
0030	Flagstone, strip veneer, random lengths				
0040	1" to 2-1/2" high	S.F.	5.19	5.06	10.25
0050	2" to 4" high	"	4.24	5.06	9.30
0060	5" to 6" high	"	3.73	5.06	8.79
0070	7" to 8" high	"	3.11	5.06	8.17

DIVISION # 04 MASONRY - QUICK ESTIMATING

			UNIT	COST
04999.10	**MASONRY**			
1000	BRICK MASONRY			
1100	Conventional (Modular Size)			
1110	Running Bond - 8" x 2 2/3" x 4"		S.F.	16.78
1120	Common Bond - 6 Course Header		"	19.59
1130	Stack Bond		"	17.04
1140	Dutch & English Bond - Every Other Course Header		"	25.00
1150	Every Course Header		"	27.75
1160	Flemish Bond - Every Other Course Header		"	18.85
1170	Every Course Header		"	22.60
1200	Add: If Scaffold Needed		"	0.84
1210	Add: For each 10' of Floor Hgt. or Floor (3%)		"	0.47
1220	Add: Piers and Corbels (15% to Labor)		"	2.39
1230	Add: Sills and Soldiers (20% to Labor)		"	2.30
1240	Add: Floor Brick (10% to Labor)		"	1.67
1300	Add: Weave & Herringbone Patterns (20% to Labor)		"	2.28
1310	Add: Stack Bond (8% to Labor)		"	1.33
1320	Add: Circular or Radius Work 20% to Labor)		"	3.49
1330	Add: Rock Faced & Slurried Face (10% to Labor)		"	1.67
1340	Add: Arches (75% to Labor)		"	5.14
1350	Add: For Winter Work (below 40°)		"	1.67
1400	Production Loss (10% to Labor)		"	1.67
1410	Enclosures - Wall Area Conventional		"	1.67
1420	Heat and Fuel - Wall Area Conventional		"	0.34
1500	Econo - 8" x 4" x 3"		"	13.84
1600	Panel - 8" x 8" x 4"		"	10.86
1700	Norman - 12" x 2 2/3" x 4"		"	13.86
1800	King Size - 10" x 2 5/8" x 4"		"	12.05
1900	Norwegian - 12" x 3 1/5" x 4"		"	11.60
2000	Saxon-Utility - 12" x 4" x 3"		"	10.88
2010	12" x 4" x 4"		"	10.93
2020	12" x 4" x 6"		"	13.79
2030	12" x 4" x 8"		"	16.46
2200	Adobe - 12" x 3" x 4"		"	18.72
2300	COATED BRICK (Ceramic)		"	22.46
2400	COMMON BRICK (Clay, Concrete and Sand Lime)		"	12.82
2500	FIRE BRICK -Light Duty		"	21.12
2510	Heavy Duty		"	26.75
2520	Deduct for Residential Work - All Above - 5%			
04999.20	**CONCRETE BLOCK**			
1000	CONVENTIONAL (Struck 2 Sides - Partitions)			
1010	12" x 8"x 16" Plain		S.F.	8.87
1020	Bond Beam (with Fill-Reinforcing)		"	11.74
1030	12" x 8"x 8" Half Block		"	10.53
1040	Double End - Header		"	9.09
1050	8" x 8" x 16" Plain		"	7.74
1060	Bond Beam (with Fill-Reinforcing)		"	9.50
1070	8" x 8" x 8" Half Block		"	7.58
1080	Double End - Header		"	8.05
1090	6 "x 8" x 16" Plain		"	7.22
2000	Bond Beam (with Fill-Reinforcing)		"	8.10
2010	Half Block		"	7.12

© 2008 By Design & Construction Resources

04A - 13

		UNIT	COST
04999.20	**CONCRETE BLOCK, Cont'd...**		
2020	4" x 8" x 16" Plain	S.F.	6.99
2030	Half Block	"	6.10
2040	16" x 8" x 16" Plain	"	10.06
2050	Half Block	"	9.59
2060	14" x 8" x 16" Plain	"	9.55
2070	Half Block	"	9.07
2080	10" x 8" x 16" Plain	"	8.27
3000	Bond Beam (with Fill-Reinforcing)	"	9.53
3010	Half Block	"	9.26
3020	Deduct: Block Struck or Cleaned One Side	"	0.29
3030	Deduct: Block Not Struck or Cleaned Two Sides	"	0.46
3040	Deduct: Clean One Side Only	"	0.20
3050	Deduct: Lt. Wt. Block (Labor Only)	"	0.19
4000	Add: Jamb and Sash Block	"	0.50
4010	Add: Bullnose Block	"	0.72
4020	Add: Pilaster, Pier, Pedestal Work	"	0.79
4030	Add: Stack Bond Work	"	0.30
4040	Add: Radius or Circular Work	"	1.50
5000	Add: If Scaffold Needed	"	0.83
5010	Add: Winter Production Cost (10% Labor)	"	0.48
5020	Add: Winter Enclosing - Wall Area	"	0.48
5030	Add: Winter Heating - Wall Area	"	0.44
5040	Add: Core & Beam Filling		
5050	Add: Insulation		
6000	Deduct for Residential Work - 5%		
8010	SCREEN WALL - 4" x 12" x 12"	S.F.	6.91
8020	BURNISHED - 12" x 8" x 16"	"	12.94
8030	8" x 8" x 16"	"	12.06
8040	6" x 8" x 16"	"	11.02
8050	4" x 8" x 16"	"	10.33
8060	2" x 8" x 16"	"	9.31
8070	Add for Shapes	"	2.80
8080	Add for 2 Faced Finish	"	4.04
8090	Add for Scored Finish - 1 Face	"	0.98
8100	PREFACED UNITS -12" x 8" x 16" Stretcher	"	18.12
8110	(Ceramic Glazed) 12" x 8" x 16" Glazed 2 Face	"	24.25
8120	8" x 8" x 16" Stretcher	"	16.92
8130	Glazed 2 Face	"	23.01
8140	6" x 8" x 16" Stretcher	"	15.12
8150	Glazed 2 Face	"	22.94
8160	4" x 8" x 16" Stretcher	"	15.86
8170	Glazed 2 Face	"	21.99
8180	2" x 8" x 16" Stretcher	"	15.73
8190	4" x 16" x 16" Stretcher	"	40.50
8200	Add for Base, Caps, Jambs, Headers, Lintels	"	4.54
8210	Add for Scored Block	"	1.34
04999.30	**CLAY BACKING AND PARTITION TILE**		
0010	3" x 12" x 12"	S.F.	6.03
0020	4" x 12" x 12"	"	6.46
0030	6" x 12" x 12"	"	7.69
0040	8" x 12" x 12"	"	8.43

		UNIT	COST
04999.40	**CLAY FACING TILE (GLAZED)**		
1000	6T or 5" x 12" - SERIES		
1010	2" x 5 1/3" x 12" Soap Stretcher (Solid Back)	S.F.	21.99
1020	4" x 5 1/2" x 12" 1 Face Stretcher	"	23.62
1030	2 Face Stretcher	"	27.75
1040	6" x 5 1/3" x 12" 1 Face Stretcher	"	28.25
1050	8" x 5 1/3" x 12" 1 Face Stretcher	"	16.18
2000	8W or 8" x 16" - SERIES		
2010	2" x 8" x 16" Soap Stretcher (Solid Back)	S.F.	16.83
2020	4" x 8" x 16" 1 Face Stretcher	"	17.22
2030	2 Face Stretcher	"	22.71
2040	6" x 8" x 16" 1 Face Stretcher	"	20.32
2050	8" x 8" x 16" 1 Face Stretcher	"	21.47
2060	Add for Shapes (Average)	PCT.	50.00
2070	Add for Designer Colors	"	20.00
2080	Add for Less than Truckload Lots	"	10.00
2090	Add for Base Only	"	25.00
04999.50	**GLASS UNITS**		
0010	4" x 8" x 4"	S.F.	55.75
0020	6" x 6" x 4"	"	59.50
0030	8" x 8" x 4"	"	36.25
0040	12" x 12" x 4"	"	36.75
04999.60	**TERRA-COTTA**		
0010	Unglazed	S.F.	12.55
0020	Glazed	"	16.92
0030	Colored Glazed	"	20.32
04999.70	**FLUE LINING**		
0010	8" x 12"	S.F.	20.84
0020	12" x 12"	"	22.62
0030	16" x 16"	"	36.25
0040	18" x 18"	"	37.00
0050	20" x 20"	"	72.75
0060	24" x 24"	"	92.50
04999.80	**NATURAL STONE**		
1000	CUT STONE BY S.F.		
1010	Limestone- Indiana and Alabama - 3"	S.F.	41.75
1020	4"	"	46.00
1030	Minnesota, Wisc, Texas, etc.- 3"	"	46.75
1040	4"	"	53.50
1200	Marble - 2"	"	57.75
1210	3"	"	61.50
1300	Granite - 2"	"	54.00
1310	3"	"	58.25
1400	Slate - 1 1/2"	"	57.75
2000	Ashlar - 4" Sawed Bed		
2100	Limestone- Indiana - Random	S.F.	32.50
2110	Coursed - 2" - 5" - 8"	"	31.25
2120	Minnesota, Alabama, Wisconsin, etc.		
2130	Split Face - Random	S.F.	32.50
2140	Coursed	"	37.00
2150	Sawed or Planed Face - Random	"	35.25
2160	Coursed	"	37.50

		UNIT	COST
04999.80	**NATURAL STONE, Cont'd...**		
2200	Marble - Sawed - Random	S.F.	46.50
2210	Coursed	"	47.25
2300	Granite - Bushhammered - Random	"	47.50
2310	Coursed	"	52.50
2400	Quartzite	"	36.50
3000	ROUGH STONE		
3100	Rubble and Flagstone	S.F.	33.50
3200	Field Stone or Boulders	"	31.25
3300	Light Weight Boulders (Igneous)		
3310	2" to 4" Veneer - Sawed Back	S.F.	27.00
3320	3" to 10" Boulders	"	28.50
3330	CUT STONE BY C.F		
3400	Limestone- Indiana and Alabama - 3"	C.F.	187
3410	4"	"	132
3420	Minnesota, Wisc, Texas, etc.- 3"	"	196
3430	4"	"	184
3500	Marble - 2"	"	400
3510	3"	"	250
3600	Granite - 2"	"	370
3610	3"	"	250
3700	Slate - 1 1/2"	"	380
3800	ASHLAR Limestone- Indiana - Random	"	97.25
3810	Coursed - 2" - 5" - 8"	"	96.75
3820	Split Face - Random	"	98.25
3830	Coursed	"	120
3840	Sawed or Planed Face - Random	"	119
3850	Coursed	"	121
3900	Marble - Sawed - Random	"	149
3910	Coursed	"	150
4000	Granite - Bushhammered - Random	"	152
4010	Coursed	"	156
4100	Quartzite	"	119
4200	ROUGH STONE, Rubble and Flagstone	"	98.25
4300	Field Stone or Boulders	"	96.75
04999.90	**PRECAST VENEERS AND SIMULATED MASONRY**		
1000	ARCHITECTURAL PRECAST STONE		
1005	Limestone	S.F.	41.00
1010	Marble	"	49.00
2010	PRECAST CONCRETE	"	33.50
3010	MOSAIC GRANITE PANELS	"	46.25
04999.91	**MORTAR**		
0010	Portland Cement and Lime Mortar - Labor in Unit Costs	S.F.	122
0020	Masonry Cement Mortar - Labor in Unit Costs	"	117
0030	Mortar for Standard Brick - Material Only	"	0.51
04999.92	**CORE FILLING FOR REINFORCED CONCRETE**		
1010	Add to Block Prices		
1020	12" x 8" x 16" Plain (includes #4 Rod 16" O.C. Vertical)	S.F.	3.82
1030	Bond Beam (includes 2 #4 Rods)	"	4.51
1040	8" x 8" x 6" Plain (includes #4 Rod 16" O.C. Vertical)	"	2.33
1050	Bond Beam (includes 2 #4 Rods)	"	1.90
1060	6" x 8" x 16" Plain	"	2.06

		UNIT	COST
04999.92	**CORE FILLING FOR REINFORCED CONCRETE BLOCK, Cont'd...**		
1070	Bond Beam (includes 1 #4 Rod)	S.F.	1.06
1080	10" x 8" x 16" Bond Beam (includes 2 #4 Rods)	"	4.09
1090	14" x 8" x 16" Bond Beam (includes 2 #4 Rods)	"	4.67
04999.93	**CORE AND CAVITY FILL FOR INSULATED**		
2100	LOOSE		
2110	Core Fill - Expanded Styrene		
2120	12" x 8" x 16" Concrete Block	S.F.	1.44
2130	10" x 8" x 16" Concrete Block	"	1.27
2140	8" x 8" x 16" Concrete Block	"	0.91
2150	6" x 8" x 16" Concrete Block	"	0.79
2160	8" Brick - Jumbo - Thru the Wall	"	0.69
2170	6" Brick - Jumbo - Thru the Wall	"	0.56
2180	4" Brick - Jumbo	"	0.48
2190	Cavity Fill - per Inch		
2200	Expanded Styrene	S.F.	0.62
2210	Mica	"	1.03
2220	Fiber Glass	"	0.63
2230	Rock Wool	"	0.63
2240	Cellulose	"	0.62
2300	RIGID - Fiber Glass - 3# Density, 1"	"	1.06
2310	2"1.58 -		
2320	Expanded Styrene - Molded - 1# Density, 1"	S.F.	0.78
2330	2"	"	1.17
2340	Extruded - 2# Density, 1"	"	1.13
2350	2"	"	1.70
2360	Expanded Urethane - 1"	"	1.21
2370	2"	"	1.60
2380	Perlite - 1"	"	1.17
2390	2"	"	1.81
2400	Add for Embedded Water Type	"	1.81
2410	Add for Glued Applications	"	0.07
04999.95	**CLEANING AND POINTING**		
4010	Brick and Stone - Acid or Chemicals	S.F.	0.86
4020	Soap and Water	"	0.81
4030	Block & Facing Tile - 1 Face (Incl. Hollow Metal Frames)	"	0.45
4040	Point with White Cement	"	1.71
04999.97	**SCAFFOLD**		
7100	EQUIPMENT, TOOLS AND BLADES		
7110	Percentage of Labor as an average	PCT.	7.00
7120	See Division 1 for Rental Rates & New Costs		
7200	SCAFFOLD- Tubular Frame - to 40 feet - Exterior	S.F.	1.08
7210	Tubular Frame - to 16 feet - Interior	"	1.03
7220	Swing Stage - 40 feet and up	"	1.14

Design & Construction Resources

TABLE OF CONTENTS   PAGE

05050.10 - STRUCTURAL WELDING	05-2
05050.30 - MECHANICAL ANCHORS	05-2
05050.90 - METAL ANCHORS	05-4
05050.95 - METAL LINTELS	05-4
05120.10 - STRUCTURAL STEEL	05-4
05410.10 - METAL FRAMING	05-4
05510.10 - STAIRS	05-5
05515.10 - LADDERS	05-6
05520.10 - RAILINGS	05-6
05580.10 - METAL SPECIALTIES	05-6
05700.10 - ORNAMENTAL METAL	05-7
QUICK ESTIMATING	**05A-8**

		UNIT	LABOR	MAT.	TOTAL
05050.10	**STRUCTURAL WELDING**				
0080	Welding				
0100	Single pass				
0120	1/8"	L.F.	2.68	0.33	3.01
0140	3/16"	"	3.58	0.55	4.13
0160	1/4"	"	4.47	0.77	5.24
05050.30	**MECHANICAL ANCHORS**				
0010	Drop-in, stainless steel, for masonry, 3/8"	EA.			0.97
0020	1/2"	"			1.07
0030	Hollow wall, for use in gypsum drywall, 6-32"x				
0040	1-1/4"	EA.			0.33
0050	2-3/8"	"			0.38
0060	1-3/4"	"			0.28
0070	1-1/2"	"			0.27
0080	2"	"			0.27
1000	Concrete anchor, 3/16"x				
1010	1-1/4"	EA.			0.44
1020	1-3/4"	"			0.55
1030	2-3/4"	"			0.55
1040	1/4"x				
1050	1-1/4"	EA.			0.55
1060	2-1/4"	"			0.66
1070	2-3/4"	"			0.71
1080	Toggle anchor, 5/8"	"			1.50
2000	3/4"	"			1.50
2010	Toggle bolts, 1/8"x				
2020	2"	EA.			0.27
2030	3"	"			0.27
2040	4"	"			0.27
2050	3/16"x				
2060	2"	EA.			0.55
2070	3"	"			0.55
2080	4"	"			0.55
2090	1/4"x				
3000	2"	EA.			0.80
3010	3"	"			0.80
3020	4"	"			0.80
3030	6"	"			1.61
3050	Sleeve anchor, 3/8"x				
3060	1-7/8"	EA.			0.34
3070	2-1/4"	"			0.38
3080	3"	"			0.44
3090	1/4x2-1/4"	"			0.39
4000	1/2"x				
4010	1/4"	EA.			0.44
4020	3"	"			0.49
4030	4"	"			0.66
4040	5/8"x				
4050	1/4"	EA.			0.77
4060	4-1/4"	"			0.77
4070	6"	"			0.82
4080	Machine screw, corrosion resistant, for use in masonry, 5/8"x				

		UNIT	LABOR	MAT.	TOTAL
05050.30	**MECHANICAL ANCHORS, Cont'd...**				
4090	2"	EA.			0.72
5000	4"	"			1.10
5010	6"	"			1.85
5020	8"	"			2.69
5030	Wedge anchor, 1/4"x2-1/4"	"			0.38
5040	3/8"x				
5050	2-1/4"	EA.			0.38
5060	3"	"			0.49
5070	3/4"	"			0.55
5080	5"	"			0.66
5090	1/2"x				
6000	3/4"	EA.			0.52
6010	3-1/4"	"			0.55
6020	4-1/4"	"			0.66
6030	5-1/2"	"			0.82
6040	5/8"x				
6050	6"	EA.			1.10
6060	7"	"			1.21
6070	8"	"			1.54
6080	Spring wing, toggle bolt, for use in hollow walls, 1/8"x				
6090	2"	EA.			0.39
7000	3"	"			0.39
7010	4"	"			0.39
7020	3/16"x				
7030	2"	EA.			0.55
7040	3"	"			0.55
7050	4"	"			0.55
7060	5"	"			0.55
7070	1/4"x				
7080	2"	EA.			0.80
7090	3"	"			0.80
8000	4"	"			0.80
8010	3/8"x6"	"			1.61
8020	Stud anchor, 1-1/4"	"			0.55
8030	3/4"	"			0.55
8040	Hex bolt, zinc plated, 3/4"x				
8050	4"	EA.			2.05
8060	6"	"			2.51
8070	8"	"			3.00
8080	1/2"x				
8090	4"	EA.			0.85
9000	6"	"			1.33
9010	8"	"			1.69
9020	Lag screw, 1/4"x				
9030	2"	EA.			0.17
9040	4"	"			0.38
9050	6"	"			0.61
9060	Machine screw, zinc plated, 5/8"x				
9070	4"	EA.			1.10
9100	6"	"			1.85
9110	8"	"			2.69
9120	Ribbed plastic anchor 5/8"x				

		UNIT	LABOR	MAT.	TOTAL
05050.30	**MECHANICAL ANCHORS, Cont'd...**				
9130	1-1/4"	EA.			0.23
9140	1-1/2"	"			0.33
05050.90	**METAL ANCHORS**				
1000	Anchor bolts				
1020	3/8" x				
1040	8" long	EA.			0.97
1080	12" long	"			1.16
1090	1/2" x				
1100	8" long	EA.			1.46
1140	12" long	"			1.70
1170	5/8" x				
1180	8" long	EA.			1.36
1220	12" long	"			1.61
1270	3/4" x				
1280	8" long	EA.			1.94
1300	12" long	"			2.19
4480	Non-drilling anchor				
4500	1/4"	EA.			0.64
4540	3/8"	"			0.80
4560	1/2"	"			1.23
7000	Self-drilling anchor				
7020	1/4"	EA.			1.62
7060	3/8"	"			2.44
7080	1/2"	"			3.25
05050.95	**METAL LINTELS**				
0080	Lintels, steel				
0100	Plain	Lb.	1.34	1.21	2.55
0120	Galvanized	"	1.34	1.82	3.16
05120.10	**STRUCTURAL STEEL**				
0100	Beams and girders, A-36				
0120	Welded	TON	630	2,410	3,040
0140	Bolted	"	580	2,360	2,940
0180	Columns				
0185	Pipe				
0190	6" dia.	Lb.	0.63	1.36	1.99
1300	Structural tube				
1310	6" square				
1320	Light sections	TON	1,270	3,170	4,440
05410.10	**METAL FRAMING**				
0100	Furring channel, galvanized				
0110	Beams and columns, 3/4"				
0120	12" o.c.	S.F.	5.37	0.51	5.88
0140	16" o.c.	"	4.88	0.39	5.27
0150	Walls, 3/4"				
0160	12" o.c.	S.F.	2.68	0.51	3.19
0170	16" o.c.	"	2.23	0.39	2.62
0172	24" o.c.	"	1.79	0.26	2.05
0173	1-1/2"				
0174	12" o.c.	S.F.	2.68	0.66	3.34
0175	16" o.c.	"	2.23	0.49	2.72
0176	24" o.c.	"	1.79	0.34	2.13

		UNIT	LABOR	MAT.	TOTAL

05410.10 METAL FRAMING, Cont'd...

		UNIT	LABOR	MAT.	TOTAL
0177	Stud, load bearing				
0178	16" o.c.				
0179	16 ga.				
0180	2-1/2"	S.F.	2.38	1.21	3.59
0190	3-5/8"	"	2.38	1.43	3.81
0200	4"	"	2.38	1.48	3.86
0220	6"	"	2.68	1.87	4.55
0280	18 ga.				
0300	2-1/2"	S.F.	2.38	0.99	3.37
0310	3-5/8"	"	2.38	1.21	3.59
0320	4"	"	2.38	1.26	3.64
0330	6"	"	2.68	1.59	4.27
0350	8"	"	2.68	1.92	4.60
0360	20 ga.				
0370	2-1/2"	S.F.	2.38	0.55	2.93
0390	3-5/8"	"	2.38	0.66	3.04
0400	4"	"	2.38	0.71	3.09
0420	6"	"	2.68	0.88	3.56
0440	8"	"	2.68	1.04	3.72
0480	24" o.c.				
0490	16 ga.				
0500	2-1/2"	S.F.	2.06	0.82	2.88
0510	3-5/8"	"	2.06	0.99	3.05
0520	4"	"	2.06	1.04	3.10
0530	6"	"	2.23	1.26	3.49
0540	8"	"	2.23	1.59	3.82
0545	18 ga.				
0550	2-1/2"	S.F.	2.06	0.66	2.72
0560	3-5/8"	"	2.06	0.77	2.83
0570	4"	"	2.06	0.82	2.88
0580	6"	"	2.23	1.04	3.27
0590	8"	"	2.23	1.26	3.49
0595	20 ga.				
0600	2-1/2"	S.F.	2.06	0.44	2.50
0610	3-5/8"	"	2.06	0.49	2.55
0620	4"	"	2.06	0.55	2.61
0630	6"	"	2.23	0.71	2.94
0640	8"	"	2.23	0.88	3.11

05510.10 STAIRS

		UNIT	LABOR	MAT.	TOTAL
1000	Stock unit, steel, complete, per riser				
1010	Tread				
1020	3'-6" wide	EA.	67.00	130	197
1040	4' wide	"	77.00	150	227
1060	5' wide	"	90.00	170	260
1200	Metal pan stair, cement filled, per riser				
1220	3'-6" wide	EA.	54.00	140	194
1240	4' wide	"	60.00	160	220
1260	5' wide	"	67.00	180	247
1280	Landing, steel pan	S.F.	13.50	55.00	68.50
1300	Cast iron tread, steel stringers, stock units, per riser				
1310	Tread				

		UNIT	LABOR	MAT.	TOTAL
05510.10	**STAIRS, Cont'd...**				
1320	3'-6" wide	EA.	67.00	250	317
1340	4' wide	"	77.00	280	357
1360	5' wide	"	90.00	340	430
1400	Stair treads, abrasive, 12" x 3'-6"				
1410	Cast iron				
1420	3/8"	EA.	26.75	140	167
1440	1/2"	"	26.75	170	197
1450	Cast aluminum				
1460	5/16"	EA.	26.75	160	187
1480	3/8"	"	26.75	170	197
1500	1/2"	"	26.75	200	227
05515.10	**LADDERS**				
0100	Ladder, 18" wide				
0110	With cage	L.F.	35.75	94.00	130
0120	Without cage	"	26.75	59.00	85.75
05520.10	**RAILINGS**				
0080	Railing, pipe				
0090	1-1/4" diameter, welded steel				
0095	2-rail				
0100	Primed	L.F.	10.75	19.75	30.50
0120	Galvanized	"	10.75	25.25	36.00
0130	3-rail				
0140	Primed	L.F.	13.50	25.25	38.75
0160	Galvanized	"	13.50	33.00	46.50
0170	Wall mounted, single rail, welded steel				
0180	Primed	L.F.	8.26	13.25	21.51
0200	Galvanized	"	8.26	17.25	25.51
0210	1-1/2" diameter, welded steel				
0215	2-rail				
0220	Primed	L.F.	10.75	21.50	32.25
0240	Galvanized	"	10.75	28.00	38.75
0245	3-rail				
0250	Primed	L.F.	13.50	27.00	40.50
0260	Galvanized	"	13.50	35.00	48.50
0270	Wall mounted, single rail, welded steel				
0280	Primed	L.F.	8.26	14.25	22.51
0300	Galvanized	"	8.26	18.50	26.76
0960	2" diameter, welded steel				
0980	2-rail				
1000	Primed	L.F.	12.00	27.00	39.00
1020	Galvanized	"	12.00	35.00	47.00
1030	3-rail				
1040	Primed	L.F.	15.25	34.00	49.25
1070	Galvanized	"	15.25	44.25	59.50
1075	Wall mounted, single rail, welded steel				
1080	Primed	L.F.	8.95	15.50	24.45
1100	Galvanized	"	8.95	20.25	29.20
05580.10	**METAL SPECIALTIES**				
0060	Kick plate				
0080	4" high x 1/4" thick				
0100	Primed	L.F.	10.75	7.31	18.06

		UNIT	LABOR	MAT.	TOTAL
05580.10	**METAL SPECIALTIES, Cont'd...**				
0120	Galvanized	L.F.	10.75	8.30	19.05
0130	6" high x 1/4" thick				
0140	Primed	L.F.	12.00	7.92	19.92
0160	Galvanized	"	12.00	9.40	21.40
0200	Fire Escape (10'-12' high)				
0210	Landing with fixed stair, 3'-6" wide	EA.	1,070	5,060	6,130
0220	4'-6" wide	"	1,070	5,720	6,790
05700.10	**ORNAMENTAL METAL**				
1020	Railings, vertical square bars, 6" o.c., with shaped top rails				
1040	Steel	L.F.	26.75	83.00	110
1060	Aluminum	"	26.75	110	137
1080	Bronze	"	35.75	180	216
1100	Stainless steel	"	35.75	160	196
1200	Laminated metal or wood handrails with metal supports				
1220	2-1/2" round or oval shape	L.F.	26.75	250	277
2020	Grilles and louvers				
2040	Fixed type louvers				
2060	4 through 10 sf	S.F.	8.95	28.25	37.20
2080	Over 10 sf	"	6.71	33.25	39.96
2200	Movable type louvers				
2220	4 through 10 sf	S.F.	8.95	33.25	42.20
2240	Over 10 sf	"	6.71	36.75	43.46
3000	Aluminum louvers				
3010	Residential use, fixed type, with screen				
3020	8" x 8"	EA.	26.75	17.25	44.00
3060	12" x 12"	"	26.75	19.00	45.75
3080	12" x 18"	"	26.75	22.75	49.50
3100	14" x 24"	"	26.75	32.75	59.50
3120	18" x 24"	"	26.75	36.75	63.50
3140	30" x 24"	"	29.75	50.00	79.75

		UNIT	COST
05999.10	**STRUCTURAL STEEL FRAME**		
1000	To 30 Ton 20' Span	S.F.	11.07
1010	24' Span	"	11.67
1020	28' Span	"	11.80
1030	32' Span	"	14.15
1040	36' Span	"	15.77
2010	44' Span	"	17.98
2020	48' Span	"	19.31
2030	52' Span	"	21.55
2050	60' Span	"	25.06
3000	Deduct for Over 30 Ton	PCT.	5.00
05999.40	**OPEN WEB JOISTS**		
1000	To 20 Ton, 20' Span	S.F.	4.08
1010	24' Span	"	4.12
1030	32' Span	"	4.20
2000	40' Span	"	4.85
2010	48' Span	"	5.74
2030	56' Span	"	6.71
2040	60' Span	"	7.20
3000	Deduct for Over 20 Ton	PCT.	10.00
05999.50	**METAL DECKING**		
1000	1/2" Deep - Ribbed - Baked Enamel - 18 Ga	S.F.	4.36
1020	22 Ga	"	3.68
2000	3" Deep - Ribbed - Baked Enamel - 18 Ga	"	8.69
2020	22 Ga	"	7.65
3000	4 1/2" Deep - Ribbed - Baked Enamel - 16 Ga	"	11.43
3010	18 Ga	"	9.88
3020	20 Ga	"	9.20
4000	3" Deep - Cellular - 18 Ga	"	11.15
4010	16 Ga	"	13.76
4020	4 1/2" Deep - Cellular - 18 Ga	"	15.77
4030	16 Ga	"	18.06
4040	Add for Galvanized	PCT	12.00
4050	Corrugated Black Standard .015	S.F.	2.23
4060	Heavy Duty - 26 Ga	"	2.41
4070	S. Duty - 24 Ga	"	3.16
4080	22 Ga	"	3.27
5000	Add for Galvanized	PCT.	15.00
05999.60	**METAL SIDING AND ROOFING**		
1000	Aluminum- Anodized	S.F.	5.46
1010	Porcelainized	"	10.87
1020	Corrugated	"	3.70
1030	Enamel, Baked Ribbed	"	5.80
1040	24 Ga Ribbed	"	3.96
1050	Acrylic	"	6.78
2000	Porcelain Ribbed	"	7.48
2010	Galvanized- Corrugated	"	3.74
2020	Plastic Faced	"	8.12
2030	Protected Metal	"	9.90
2040	Add for Liner Panels	"	3.08
2050	Add for Insulation	"	0.46

TABLE OF CONTENTS   PAGE

06110.10 - BLOCKING	**06-2**
06110.20 - CEILING FRAMING	06-2
06110.30 - FLOOR FRAMING	**06-2**
06110.40 - FURRING	**06-3**
06110.50 - ROOF FRAMING	**06-3**
06110.60 - SLEEPERS	06-5
06110.65 - SOFFITS	**06-5**
06110.70 - WALL FRAMING	06-5
06115.10 - FLOOR SHEATHING	**06-6**
06115.20 - ROOF SHEATHING	06-6
06115.30 - WALL SHEATHING	**06-6**
06125.10 - WOOD DECKING	06-7
06130.10 - HEAVY TIMBER	**06-7**
06190.20 - WOOD TRUSSES	06-8
06190.30 - LAMINATED BEAMS	**06-8**
06200.10 - FINISH CARPENTRY	06-9
06220.10 - MILLWORK	**06-11**
06300.10 - WOOD TREATMENT	06-11
06430.10 - STAIRWORK	**06-12**
06440.10 - COLUMNS	06-12
QUICK ESTIMATING	**06A-13**

		UNIT	LABOR	MAT.	TOTAL
06110.10	**BLOCKING**				
1215	Wood construction				
1220	Walls				
1230	2x4	L.F.	2.68	0.62	3.30
1240	2x6	"	3.01	0.88	3.89
1250	2x8	"	3.21	1.26	4.47
1260	2x10	"	3.44	1.73	5.17
1270	2x12	"	3.71	2.29	6.00
1280	Ceilings				
1290	2x4	L.F.	3.01	0.62	3.63
1300	2x6	"	3.44	0.88	4.32
1310	2x8	"	3.71	1.26	4.97
1320	2x10	"	4.02	1.73	5.75
1330	2x12	"	4.38	2.29	6.67
06110.20	**CEILING FRAMING**				
1000	Ceiling joists				
1070	16" o.c.				
1080	2x4	S.F.	0.92	0.62	1.54
1090	2x6	"	0.96	0.94	1.90
1100	2x8	"	1.00	1.32	2.32
1110	2x10	"	1.04	1.95	2.99
1120	2x12	"	1.09	2.53	3.62
1130	24" o.c.				
1140	2x4	S.F.	0.76	0.49	1.25
1150	2x6	"	0.80	0.75	1.55
1160	2x8	"	0.84	1.01	1.85
1170	2x10	"	0.89	1.55	2.44
1180	2x12	"	0.94	2.09	3.03
1200	Headers and nailers				
1210	2x4	L.F.	1.55	0.62	2.17
1220	2x6	"	1.60	0.88	2.48
1230	2x8	"	1.72	1.26	2.98
1240	2x10	"	1.85	1.73	3.58
1250	2x12	"	2.01	2.29	4.30
1300	Sister joists for ceilings				
1310	2x4	L.F.	3.44	0.62	4.06
1320	2x6	"	4.02	0.88	4.90
1330	2x8	"	4.82	1.26	6.08
1340	2x10	"	6.03	1.73	7.76
1350	2x12	"	8.04	2.29	10.33
06110.30	**FLOOR FRAMING**				
1000	Floor joists				
1180	16" o.c.				
1190	2x6	S.F.	0.80	0.94	1.74
1200	2x8	"	0.81	1.32	2.13
1220	2x10	"	0.83	1.95	2.78
1230	2x12	"	0.86	2.53	3.39
1240	2x14	"	0.89	3.12	4.01
1250	3x6	"	0.83	2.12	2.95
1260	3x8	"	0.86	2.76	3.62
1270	3x10	"	0.89	3.44	4.33
1280	3x12	"	0.92	4.11	5.03

		UNIT	LABOR	MAT.	TOTAL
06110.30	**FLOOR FRAMING, Cont'd...**				
1290	3x14	S.F.	0.96	4.84	5.80
1300	4x6	"	0.83	2.75	3.58
1310	4x8	"	0.86	3.68	4.54
1320	4x10	"	0.89	4.58	5.47
1330	4x12	"	0.92	5.42	6.34
1340	4x14	"	0.96	6.43	7.39
2000	Sister joists for floors				
2010	2x4	L.F.	3.01	0.62	3.63
2020	2x6	"	3.44	0.88	4.32
2030	2x8	"	4.02	1.26	5.28
2040	2x10	"	4.82	1.73	6.55
2050	2x12	"	6.03	2.29	8.32
2060	3x6	"	4.82	2.12	6.94
2070	3x8	"	5.36	2.76	8.12
2080	3x10	"	6.03	3.44	9.47
2090	3x12	"	6.89	4.11	11.00
2100	4x6	"	4.82	2.67	7.49
2110	4x8	"	5.36	3.68	9.04
2120	4x10	"	6.03	4.58	10.61
2130	4x12	"	6.89	5.42	12.31
06110.40	**FURRING**				
1100	Furring, wood strips				
1102	Walls				
1105	On masonry or concrete walls				
1107	1x2 furring				
1110	12" o.c.	S.F.	1.50	0.29	1.79
1120	16" o.c.	"	1.37	0.24	1.61
1130	24" o.c.	"	1.27	0.20	1.47
1135	1x3 furring				
1140	12" o.c.	S.F.	1.50	0.42	1.92
1150	16" o.c.	"	1.37	0.36	1.73
1160	24" o.c.	"	1.27	0.29	1.56
1165	On wood walls				
1167	1x2 furring				
1170	12" o.c.	S.F.	1.07	0.29	1.36
1180	16" o.c.	"	0.96	0.24	1.20
1190	24" o.c.	"	0.87	0.20	1.07
1195	1x3 furring				
1200	12" o.c.	S.F.	1.07	0.42	1.49
1210	16" o.c.	"	0.96	0.36	1.32
1220	24" o.c.	"	0.87	0.29	1.16
1224	Ceilings				
06110.50	**ROOF FRAMING**				
1000	Roof framing				
1005	Rafters, gable end				
1008	0-2 pitch (flat to 2-in-12)				
1070	16" o.c.				
1080	2x6	S.F.	0.86	0.94	1.80
1090	2x8	"	0.89	1.32	2.21
1100	2x10	"	0.92	1.96	2.88
1110	2x12	"	0.96	2.53	3.49

		UNIT	LABOR	MAT.	TOTAL
06110.50	**ROOF FRAMING, Cont'd...**				
1120	24" o.c.				
1130	2x6	S.F.	0.73	0.75	1.48
1140	2x8	"	0.75	1.01	1.76
1150	2x10	"	0.77	1.55	2.32
1160	2x12	"	0.80	2.09	2.89
1165	4-6 pitch (4-in-12 to 6-in-12)				
1220	16" o.c.				
1230	2x6	S.F.	0.89	0.98	1.87
1240	2x8	"	0.92	1.44	2.36
1250	2x10	"	0.96	2.09	3.05
1260	2x12	"	1.00	2.80	3.80
1270	24" o.c.				
1280	2x6	S.F.	0.75	0.80	1.55
1290	2x8	"	0.77	1.10	1.87
1300	2x10	"	0.83	1.65	2.48
1310	2x12	"	0.92	2.24	3.16
1315	8-12 pitch (8-in-12 to 12-in-12)				
1380	16" o.c.				
1390	2x6	S.F.	0.92	1.07	1.99
1400	2x8	"	0.96	1.49	2.45
1410	2x10	"	1.00	2.24	3.24
1420	2x12	"	1.04	2.98	4.02
1430	24" o.c.				
1440	2x6	S.F.	0.77	0.82	1.59
1450	2x8	"	0.80	1.16	1.96
1460	2x10	"	0.83	1.77	2.60
1470	2x12	"	0.86	2.37	3.23
2000	Ridge boards				
2010	2x6	L.F.	2.41	0.88	3.29
2020	2x8	"	2.68	1.26	3.94
2030	2x10	"	3.01	1.73	4.74
2040	2x12	"	3.44	2.29	5.73
3000	Hip rafters				
3010	2x6	L.F.	1.72	0.88	2.60
3020	2x8	"	1.78	1.26	3.04
3030	2x10	"	1.85	1.73	3.58
3040	2x12	"	1.93	2.29	4.22
3180	Jack rafters				
3190	4-6 pitch (4-in-12 to 6-in-12)				
3200	16" o.c.				
3210	2x6	S.F.	1.42	1.00	2.42
3220	2x8	"	1.46	1.42	2.88
3230	2x10	"	1.55	2.09	3.64
3240	2x12	"	1.60	2.82	4.42
3250	24" o.c.				
3260	2x6	S.F.	1.09	0.80	1.89
3270	2x8	"	1.12	1.10	2.22
3280	2x10	"	1.17	1.65	2.82
3290	2x12	"	1.20	2.24	3.44
3295	8-12 pitch (8-in-12 to 12-in-12)				
3300	16" o.c.				
3310	2x6	S.F.	1.50	1.57	3.07

		UNIT	LABOR	MAT.	TOTAL
06110.50	**ROOF FRAMING, Cont'd...**				
3320	2x8	S.F.	1.55	2.21	3.76
3330	2x10	"	1.60	3.30	4.90
3340	2x12	"	1.66	4.40	6.06
3350	24" o.c.				
3360	2x6	S.F.	1.14	1.28	2.42
3370	2x8	"	1.17	1.77	2.94
3380	2x10	"	1.20	2.67	3.87
3390	2x12	"	1.23	3.53	4.76
4980	Sister rafters				
5000	2x4	L.F.	3.44	0.62	4.06
5010	2x6	"	4.02	0.88	4.90
5020	2x8	"	4.82	1.26	6.08
5030	2x10	"	6.03	1.73	7.76
5040	2x12	"	8.04	2.29	10.33
5050	Fascia boards				
5060	2x4	L.F.	2.41	0.62	3.03
5070	2x6	"	2.41	0.88	3.29
5080	2x8	"	2.68	1.26	3.94
5090	2x10	"	2.68	1.73	4.41
5100	2x12	"	3.01	2.29	5.30
7980	Cant strips				
7985	Fiber				
8000	3x3	L.F.	1.37	0.30	1.67
8020	4x4	"	1.46	0.40	1.86
8030	Wood				
8040	3x3	L.F.	1.46	1.68	3.14
06110.60	**SLEEPERS**				
0960	Sleepers, over concrete				
1090	16" o.c.				
1100	1x2	S.F.	0.96	0.21	1.17
1120	1x3	"	0.96	0.30	1.26
1140	2x4	"	1.14	0.62	1.76
1160	2x6	"	1.20	0.94	2.14
06110.65	**SOFFITS**				
0980	Soffit framing				
1000	2x3	L.F.	3.44	0.49	3.93
1020	2x4	"	3.71	0.62	4.33
1030	2x6	"	4.02	0.88	4.90
1040	2x8	"	4.38	1.26	5.64
06110.70	**WALL FRAMING**				
0960	Framing wall, studs				
1110	16" o.c.				
1120	2x3	S.F.	0.75	0.46	1.21
1140	2x4	"	0.75	0.62	1.37
1150	2x6	"	0.80	0.88	1.68
1160	2x8	"	0.83	1.32	2.15
1165	24" o.c.				
1170	2x3	S.F.	0.65	0.34	0.99
1180	2x4	"	0.65	0.50	1.15
1190	2x6	"	0.68	0.75	1.43
1200	2x8	"	0.71	1.01	1.72

		UNIT	LABOR	MAT.	TOTAL
06110.70	**WALL FRAMING, Cont'd...**				
1480	Plates, top or bottom				
1500	2x3	L.F.	1.42	0.49	1.91
1510	2x4	"	1.50	0.62	2.12
1520	2x6	"	1.60	0.88	2.48
1530	2x8	"	1.72	1.26	2.98
2000	Headers, door or window				
2044	2x8				
2046	Single				
2050	4' long	EA.	30.25	5.26	35.51
2060	8' long	"	37.25	10.50	47.75
2065	Double				
2070	4' long	EA.	34.50	10.50	45.00
2080	8' long	"	44.00	21.00	65.00
2134	2x12				
2138	Single				
2140	6' long	EA.	37.25	15.25	52.50
2150	12' long	"	48.25	30.50	78.75
2155	Double				
2160	6' long	EA.	44.00	30.50	74.50
2170	12' long	"	54.00	61.00	115
06115.10	**FLOOR SHEATHING**				
1980	Sub-flooring, plywood, CDX				
2000	1/2" thick	S.F.	0.60	0.69	1.29
2020	5/8" thick	"	0.68	0.83	1.51
2080	3/4" thick	"	0.80	0.98	1.78
2090	Structural plywood				
2100	1/2" thick	S.F.	0.60	0.77	1.37
2120	5/8" thick	"	0.68	0.90	1.58
2140	3/4" thick	"	0.74	1.07	1.81
5990	Underlayment				
6000	Hardboard, 1/4" tempered	S.F.	0.60	0.49	1.09
6010	Plywood, CDX				
6020	3/8" thick	S.F.	0.60	0.61	1.21
6040	1/2" thick	"	0.64	0.70	1.34
6060	5/8" thick	"	0.68	0.84	1.52
6080	3/4" thick	"	0.74	1.00	1.74
06115.20	**ROOF SHEATHING**				
0080	Sheathing				
0090	Plywood, CDX				
1000	3/8" thick	S.F.	0.62	0.61	1.23
1020	1/2" thick	"	0.64	0.70	1.34
1040	5/8" thick	"	0.68	0.84	1.52
1060	3/4" thick	"	0.74	1.00	1.74
1080	Structural plywood				
2040	3/8" thick	S.F.	0.62	0.67	1.29
2060	1/2" thick	"	0.64	0.77	1.41
2080	5/8" thick	"	0.68	0.92	1.60
2100	3/4" thick	"	0.74	1.10	1.84
06115.30	**WALL SHEATHING**				
0980	Sheathing				
0990	Plywood, CDX				

DIVISION # 06 WOOD AND PLASTICS

		UNIT	LABOR	MAT.	TOTAL
06115.30	**WALL SHEATHING, Cont'd...**				
1000	3/8" thick	S.F.	0.71	0.61	1.32
1020	1/2" thick	"	0.74	0.70	1.44
1040	5/8" thick	"	0.80	0.84	1.64
1060	3/4" thick	"	0.87	1.00	1.87
3000	Waferboard				
3020	3/8" thick	S.F.	0.71	0.53	1.24
3040	1/2" thick	"	0.74	0.66	1.40
3060	5/8" thick	"	0.80	0.80	1.60
3080	3/4" thick	"	0.87	0.87	1.74
4100	Structural plywood				
4120	3/8" thick	S.F.	0.71	0.67	1.38
4140	1/2" thick	"	0.74	0.77	1.51
4160	5/8" thick	"	0.80	0.92	1.72
4180	3/4" thick	"	0.87	1.10	1.97
7000	Gypsum, 1/2" thick	"	0.74	0.64	1.38
8000	Asphalt impregnated fiberboard, 1/2" thick	"	0.74	0.72	1.46
06125.10	**WOOD DECKING**				
0090	Decking, T&G solid				
0095	Cedar				
0100	3" thick	S.F.	1.20	8.29	9.49
0120	4" thick	"	1.28	10.00	11.28
1030	Fir				
1040	3" thick	S.F.	1.20	3.46	4.66
1060	4" thick	"	1.28	4.23	5.51
1080	Southern yellow pine				
2000	3" thick	S.F.	1.37	3.40	4.77
2020	4" thick	"	1.48	3.65	5.13
3120	White pine				
3140	3" thick	S.F.	1.20	4.30	5.50
3160	4" thick	"	1.28	5.68	6.96
06130.10	**HEAVY TIMBER**				
1000	Mill framed structures				
1010	Beams to 20' long				
1020	Douglas fir				
1040	6x8	L.F.	6.33	6.12	12.45
1042	6x10	"	6.55	7.22	13.77
1044	6x12	"	7.03	8.64	15.67
1046	6x14	"	7.31	10.25	17.56
1048	6x16	"	7.60	11.25	18.85
1060	8x10	"	6.55	9.58	16.13
1070	8x12	"	7.03	11.25	18.28
1080	8x14	"	7.31	13.00	20.31
1090	8x16	"	7.60	14.75	22.35
1380	Columns to 12' high				
1400	Douglas fir				
1420	6x6	L.F.	9.50	4.40	13.90
1440	8x8	"	9.50	7.54	17.04
1460	10x10	"	10.50	13.25	23.75
1480	12x12	"	10.50	16.25	26.75
0080	Posts, treated				
0100	4x4	L.F.	1.93	2.04	3.97

DIVISION # 06 WOOD AND PLASTICS

		UNIT	LABOR	MAT.	TOTAL
06132.10	**HEAVY TIMBER, Cont'd...**				
0120	6x6	L.F.	2.41	4.79	7.20
06190.20	**WOOD TRUSSES**				
0960	Truss, fink, 2x4 members				
0980	3-in-12 slope				
1030	5-in-12 slope				
1040	24' span	EA.	56.00	76.00	132
1050	28' span	"	58.00	89.00	147
1055	30' span	"	59.00	100	159
1060	32' span	"	59.00	110	169
1070	40' span	"	63.00	150	213
1074	Gable, 2x4 members				
1078	5-in-12 slope				
1080	24' span	EA.	56.00	100	156
1100	28' span	"	58.00	130	188
1120	30' span	"	59.00	130	189
1160	36' span	"	61.00	150	211
1180	40' span	"	63.00	160	223
1190	King post type, 2x4 members				
2000	4-in-12 slope				
2040	16' span	EA.	51.00	63.00	114
2060	18' span	"	53.00	67.00	120
2080	24' span	"	56.00	72.00	128
2120	30' span	"	59.00	98.00	157
2160	38' span	"	61.00	130	191
2180	42' span	"	66.00	150	216
06190.30	**LAMINATED BEAMS**				
0010	Parallel strand beams 3-1/2" wide x				
0020	9-1/2"	L.F.	2.71	7.59	10.30
0030	11-1/4"	"	2.81	8.25	11.06
0040	11-7/8"	"	2.92	9.21	12.13
0050	14"	"	3.45	11.25	14.70
0060	16"	"	3.80	13.50	17.30
0070	18"	"	4.22	16.00	20.22
1000	Laminated veneer beams, 1-3/4" wide x				
1010	11-7/8"	L.F.	2.92	4.93	7.85
1020	14"	"	3.45	5.94	9.39
1030	16"	"	3.80	6.93	10.73
1040	18"	"	4.22	8.40	12.62
2000	Laminated strand beams, 1-3/4" wide x				
2010	9-1/2"	L.F.	2.71	3.30	6.01
2020	11-7/8"	"	2.92	3.80	6.72
2030	14"	"	3.45	4.51	7.96
2040	16"	"	3.80	5.20	9.00
2050	3-1/2" wide x				
2060	9-1/2"	L.F.	2.71	6.05	8.76
2070	11-7/8"	"	2.92	7.60	10.52
2080	14"	"	3.45	8.96	12.41
2090	16"	"	3.80	10.50	14.30
3000	Gluelam beam, 3-1/2" wide x				
3010	10"	L.F.	2.71	8.80	11.51
3020	12"	"	3.16	10.25	13.41

		UNIT	LABOR	MAT.	TOTAL
06190.30	**LAMINATED BEAMS, Cont'd...**				
3030	15"	L.F.	3.62	12.75	16.37
3040	5-1/2" wide x				
3050	10"	L.F.	2.71	14.50	17.21
3060	16"	"	3.80	22.75	26.55
3070	20"	"	4.47	27.00	31.47
06200.10	**FINISH CARPENTRY**				
0070	Mouldings and trim				
0980	Apron, flat				
1000	9/16 x 2	L.F.	2.41	1.30	3.71
1010	9/16 x 3-1/2	"	2.54	2.39	4.93
1015	Base				
1020	Colonial				
1022	7/16 x 2-1/4	L.F.	2.41	1.52	3.93
1024	7/16 x 3	"	2.41	1.81	4.22
1026	7/16 x 3-1/4	"	2.41	1.98	4.39
1028	9/16 x 3	"	2.54	1.95	4.49
1030	9/16 x 3-1/4	"	2.54	2.07	4.61
1034	11/16 x 2-1/4	"	2.68	1.81	4.49
1035	Ranch				
1036	7/16 x 2-1/4	L.F.	2.41	1.57	3.98
1038	7/16 x 3-1/4	"	2.41	2.07	4.48
1039	9/16 x 2-1/4	"	2.54	1.71	4.25
1041	9/16 x 3	"	2.54	2.05	4.59
1043	9/16 x 3-1/4	"	2.54	2.07	4.61
1050	Casing				
1060	11/16 x 2-1/2	L.F.	2.19	1.58	3.77
1070	11/16 x 3-1/2	"	2.29	2.04	4.33
1180	Chair rail				
1200	9/16 x 2-1/2	L.F.	2.41	1.65	4.06
1210	9/16 x 3-1/2	"	2.41	2.48	4.89
1250	Closet pole				
1300	1-1/8" dia.	L.F.	3.21	1.89	5.10
1310	1-5/8" dia.	"	3.21	2.18	5.39
1340	Cove				
1500	9/16 x 1-3/4	L.F.	2.41	1.27	3.68
1510	11/16 x 2-3/4	"	2.41	1.75	4.16
1550	Crown				
1600	9/16 x 1-5/8	L.F.	3.21	1.90	5.11
1620	11/16 x 3-5/8	"	4.02	2.51	6.53
1640	11/16 x 5-1/4	"	4.82	4.54	9.36
1680	Drip cap				
1700	1-1/16 x 1-5/8	L.F.	2.41	1.80	4.21
1780	Glass bead				
1800	3/8 x 3/8	L.F.	3.01	0.51	3.52
1840	5/8 x 5/8	"	3.01	0.68	3.69
1860	3/4 x 3/4	"	3.01	0.84	3.85
1880	Half round				
1900	1/2	L.F.	1.93	0.63	2.56
1910	5/8	"	1.93	0.84	2.77
1920	3/4	"	1.93	1.04	2.97
1980	Lattice				

06200.10	FINISH CARPENTRY, Cont'd...	UNIT	LABOR	MAT.	TOTAL
2000	1/4 x 7/8	L.F.	1.93	0.49	2.42
2010	1/4 x 1-1/8	"	1.93	0.52	2.45
2030	1/4 x 1-3/4	"	1.93	0.65	2.58
2040	1/4 x 2	"	1.93	0.74	2.67
2080	Ogee molding				
2100	5/8 x 3/4	L.F.	2.41	1.04	3.45
2110	11/16 x 1-1/8	"	2.41	1.57	3.98
2120	11/16 x 1-3/8	"	2.41	1.89	4.30
2180	Parting bead				
2200	3/8 x 7/8	L.F.	3.01	0.81	3.82
2300	Quarter round				
2301	1/4 x 1/4	L.F.	1.93	0.24	2.17
2303	3/8 x 3/8	"	1.93	0.37	2.30
2305	1/2 x 1/2	"	1.93	0.51	2.44
2307	11/16 x 11/16	"	2.09	0.54	2.63
2309	3/4 x 3/4	"	2.09	0.60	2.69
2311	1-1/16 x 1-1/16	"	2.19	0.75	2.94
2380	Railings, balusters				
2400	1-1/8 x 1-1/8	L.F.	4.82	2.75	7.57
2410	1-1/2 x 1-1/2	"	4.38	3.22	7.60
2480	Screen moldings				
2500	1/4 x 3/4	L.F.	4.02	0.66	4.68
2510	5/8 x 5/16	"	4.02	0.83	4.85
2580	Shoe				
2600	7/16 x 11/16	L.F.	1.93	0.83	2.76
2605	Sash beads				
2620	1/2 x 7/8	L.F.	4.02	1.21	5.23
2640	5/8 x 7/8	"	4.38	1.28	5.66
2760	Stop				
2780	5/8 x 1-5/8				
2800	Colonial	L.F.	3.01	0.65	3.66
2810	Ranch	"	3.01	0.63	3.64
2880	Stools				
2900	11/16 x 2-1/4	L.F.	5.36	2.93	8.29
2910	11/16 x 2-1/2	"	5.36	3.11	8.47
2920	11/16 x 5-1/4	"	6.03	3.33	9.36
4000	Exterior trim, casing, select pine, 1x3	"	2.41	2.18	4.59
4010	Douglas fir				
4020	1x3	L.F.	2.41	1.16	3.57
4040	1x4	"	2.41	1.57	3.98
4060	1x6	"	2.68	1.90	4.58
4100	1x8	"	3.01	2.75	5.76
5000	Cornices, white pine, #2 or better				
5040	1x4	L.F.	2.41	1.07	3.48
5080	1x8	"	2.84	1.90	4.74
5120	1x12	"	3.21	3.22	6.43
8600	Shelving, pine				
8620	1x8	L.F.	3.71	2.27	5.98
8640	1x10	"	3.86	2.75	6.61
8660	1x12	"	4.02	3.60	7.62
8800	Plywood shelf, 3/4", with edge band, 12" wide	"	4.82	2.48	7.30
8840	Adjustable shelf, and rod, 12" wide				

		UNIT	LABOR	MAT.	TOTAL
06200.10	**FINISH CARPENTRY, Cont'd...**				
8860	3' to 4' long	EA.	12.00	16.75	28.75
8880	5' to 8' long	"	16.00	31.50	47.50
8900	Prefinished wood shelves with brackets and supports				
8905	8" wide				
8910	3' long	EA.	12.00	49.50	61.50
8922	4' long	"	12.00	57.00	69.00
8924	6' long	"	12.00	83.00	95.00
8930	10" wide				
8940	3' long	EA.	12.00	55.00	67.00
8942	4' long	"	12.00	78.00	90.00
8946	6' long	"	12.00	92.00	104
06220.10	**MILLWORK**				
0070	Countertop, laminated plastic				
0080	25" x 7/8" thick				
0099	Minimum	L.F.	12.00	18.00	30.00
0100	Average	"	16.00	32.25	48.25
0110	Maximum	"	19.25	54.00	73.25
0115	25" x 1-1/4" thick				
0120	Minimum	L.F.	16.00	21.50	37.50
0130	Average	"	19.25	35.75	55.00
0140	Maximum	"	24.25	57.00	81.25
0160	Add for cutouts	EA.	30.25		30.25
0165	Backsplash, 4" high, 7/8" thick	L.F.	9.65	18.00	27.65
2000	Plywood, sanded, A-C				
2020	1/4" thick	S.F.	1.60	1.26	2.86
2040	3/8" thick	"	1.72	1.39	3.11
2060	1/2" thick	"	1.85	1.53	3.38
2070	A-D				
2080	1/4" thick	S.F.	1.60	1.15	2.75
2090	3/8" thick	"	1.72	1.32	3.04
2100	1/2" thick	"	1.85	1.48	3.33
2500	Base cabinets, 34-1/2" high, 24" deep, hardwood, no tops				
2540	Minimum	L.F.	19.25	68.00	87.25
2560	Average	"	24.25	120	144
2580	Maximum	"	32.25	180	212
2600	Wall cabinets				
2640	Minimum	L.F.	16.00	55.00	71.00
2660	Average	"	19.25	88.00	107
2680	Maximum	"	24.25	170	194
06300.10	**WOOD TREATMENT**				
1000	Creosote preservative treatment				
1020	8 lb/cf	B.F.			0.50
1040	10 lb/cf	"			0.65
1060	Salt preservative treatment				
1070	Oil borne				
1080	Minimum	B.F.			0.38
1100	Maximum	"			0.61
1120	Water borne				
1140	Minimum	B.F.			0.29
1150	Maximum	"			0.47
1200	Fire retardant treatment				

		UNIT	LABOR	MAT.	TOTAL
06300.10	**WOOD TREATMENT, Cont'd...**				
1220	Minimum	B.F.			0.63
1240	Maximum	"			0.76
1300	Kiln dried, softwood, add to framing costs				
1320	1" thick	B.F.			0.21
1360	3" thick	"			0.44
06430.10	**STAIRWORK**				
0080	Risers, 1x8, 42" wide				
0100	White oak	EA.	24.25	22.75	47.00
0120	Pine	"	24.25	16.00	40.25
0130	Treads, 1-1/16" x 9-1/2" x 42"				
0140	White oak	EA.	30.25	32.50	62.75
06440.10	**COLUMNS**				
0980	Column, hollow, round wood				
0990	12" diameter				
1000	10' high	EA.	70.00	660	730
1080	16' high	"	110	1,200	1,310
2000	24" diameter				
2020	16' high	EA.	110	2,730	2,840
2060	20' high	"	110	3,830	3,940
2100	24' high	"	120	4,380	4,500

		UNIT	COST
06999.10	**ROUGH CARPENTRY**		
1000	LIGHT FRAMING AND SHEATHING		
1100	Joists and Headers - Floor Area - 16" O.C.		
1110	2" x 6" Joists - with Headers & Bridging	S.F.	2.33
1120	2" x 8" Joists	"	2.99
1130	2" x 10" Joists	"	3.34
1140	2" x 12" Joists	"	4.13
1150	Add for Ceiling Joists, 2nd Floor and Above	"	0.08
1160	Add for Sloped Installation	"	0.15
1200	Studs, Plates and Framing - 8' Wall Height - 16" O.C.	"	1.50
1210	2" x 3" Stud Wall - Non-Bearing (Single Top Plate)		
1220	2" x 4" Stud Wall - Bearing (Double Top Plate)	S.F.	1.71
1230	2" x 4" Stud Wall - Non-Bearing (Single Top Plate)	"	1.60
1240	2" x 6" Stud Wall - Bearing (Double Top Plate)	"	2.12
1250	2" x 6" Stud Wall - Non-Bearing (Single Top Plate)	"	2.06
1300	Add for Stud Wall - 12" O.C.	"	0.08
1310	Deduct for Stud Wall - 24" O.C.	"	0.22
1320	Add for Bolted Plates or Sills	"	0.12
1330	Add for Each Foot Above 8'	"	0.05
1340	Add to Above for Fire Stops, Fillers and Nailers	"	0.51
1350	Add for Soffits and Suspended Framing	"	0.68
1360	Bridging - 1" x 3" Wood Diagonal	EA.	2.76
1370	2" x 8" Solid	"	3.00
1400	Rafters		
1410	2" x 4" Rafter (Incl Bracing) 3 - 12 Slope	S.F.	2.01
1420	4 - 12 Slope	"	2.01
1430	5 - 12 Slope	"	2.28
1440	6 - 12 Slope	"	2.66
1500	Add for Hip-and-Valley Type	"	0.47
1510	Add for 1' 0" Overhang - Total Area of Roof	"	0.17
1520	2" x 6" Rafter (Incl Bracing) 3-12 Slope	"	2.52
1530	4-12 Slope	"	2.56
1540	5-12 Slope	"	2.66
1550	6-12 Slope	"	2.76
1560	Add for Hip - and - Valley Type	"	0.58
1600	Stairs	EA.	510
1700	Sub Floor Sheathing (Structural)		
1710	1" x 8" and 1" x 10" #3 Pine	S.F.	1.76
1720	1/2" x 4' x 8' CD Plywood - Exterior	"	1.41
1730	5/8" x 4' x 8'	"	1.59
1740	3/4" x 4' x 8'	"	1.82
1800	Floor Sheathing (Over Sub Floor)		
1810	3/8" x 4' x 8' CD Plywood	S.F.	1.33
1820	1/2" x 4' x 8'	"	1.43
1830	5/8" x 4' x 8'	"	1.55
1840	1/2" x 4' x 8' Particle Board	"	1.14
1850	5/8" x 4' x 8'	"	1.23
1860	3/4" x 4' x 8'	"	1.46
1900	Wall Sheathing		
1910	1' x 8" and 1" x 10" - #3 Pine	S.F.	1.82
1920	3/8" x 4' x 8' CD Plywood - Exterior	"	1.38
1930	1/2" x 4' x 8'	"	1.50
1940	5/8" x 4' x 8'	"	1.67

DIVISION # 06 WOOD AND PLASTICS - QUICK ESTIMATING

		UNIT	COST
06999.10	**ROUGH CARPENTRY, Cont'd...**		
1950	3/4" x 4' x 8'	S.F.	1.94
1960	25/32" x 4' x 8' Fiber Board - Impregnated	"	1.06
1970	1" x 2' x 8' T&G Styrofoam	"	1.36
1980	2" x 2' x 8'	"	1.99
2000	Roof Sheathing - Flat Construction		
2010	1" x 6" and 1" x 8" - #3 Pine	S.F.	1.79
2020	1/2" x 4' x 8' CD Plywood - Exterior	"	1.54
2030	5/8" x 4' x 8'	"	1.68
2040	3/4" x 4' x 8'	"	1.96
2050	Add for Sloped Roof Construction (to 5-12 slope)	"	0.18
2060	Add for Steep Sloped Construction (over 5-12 slope)	"	0.31
2070	Add to Above Sheathing		
2080	AC or AD Plywood	S.F.	0.31
2090	10' Length Plywood	"	0.24
2100	HEAVY FRAMING		
2200	Columns and Beams - 16' Span Average - Floor Area	S.F.	8.00
2210	20' Span	"	8.36
2220	24' Span	"	9.71
2300	Deck 2" x 6" T&G - Fir Random Construction Grade	"	4.73
2310	3" x 6" T&G - Fir Random Construction Grade	"	6.31
2320	4" x 6"	"	8.00
2330	2" x 6" T&G - Red Cedar	"	5.97
2340	Add for D Grade Cedar	"	2.00
2350	2" x 6" T&G - Panelized Fir	"	4.32
3000	MISCELLANEOUS CARPENTRY		
3100	Blocking and Bucks (2" x 4" and 2" x 6")		
3110	Doors & Windows - Nailed to Concrete or Masonry	EA.	55.50
3120	Bolted to Concrete or Masonry	"	61.00
3130	Doors & Windows - Nailed to Concrete or Masonry	B.F.	2.66
3140	Bolted to Concrete or Masonry	"	3.71
3150	Roof Edges - Nailed to Wood	"	2.12
3160	Bolted to Concrete (Incl. bolts)	"	3.19
3200	Grounds and Furring		
3210	1" x 6" Fastened to Wood	L.F.	1.19
3220	1" x 4"	"	1.02
3230	1" x 3"	"	0.97
3240	1" x 3" Fastened to Concrete/Masonry - Nailed	"	1.54
3250	Gun Driven	"	1.43
3260	PreClipped	"	1.77
3270	2" x 2" Suspended Framing	"	1.54
3280	Not Suspended Framing	"	1.33
3300	Grounds and Furring		
3310	1" x 6" Fastened to Wood	B.F.	2.39
3320	1" x 4"	"	3.19
3330	1" x 3"	"	3.92
3340	1" x 3" Fastened to Concrete/Masonry - Nailed	"	6.00
3400	Gun Driven	"	5.72
3410	PreClipped	"	7.00
3420	2" x 2" Suspended Framing	"	4.67
3430	Not Suspended Framing	"	3.99
3440	Cant Strips		
3450	4" x 4" Treated and Nailed	S.F.	1.79

		UNIT	COST
06999.10	**ROUGH CARPENTRY, Cont'd...**		
3460	Treated and Bolted (Including Bolts)	S.F.	2.60
3470	6" x 6" Treated and Nailed	"	3.60
3480	Treated and Bolted (Including Bolts)	"	4.56
3600	Building Papers and Sealers		
3610	15" Felt	S.F.	0.21
3620	Polyethylene - 4 mil	"	0.16
3630	6 mil	"	0.18
3640	Sill Sealer	"	0.81
06999.20	**FINISH CARPENTRY**		
1000	FINISH SIDINGS AND FACING MATERIALS (Exterior) TO %		
1100	Boards and Beveled Sidings		
1200	Cedar Beveled - Clear Heart 1/2" x 4"	S.F.	6.26
1210	1/2" x 6"	"	5.43
1220	1/2" x 8"	"	4.54
1300	Rough Sawn 7/8" x 8"	"	4.67
1310	7/8" x 10"	"	4.84
1320	7/8" x 12"	"	4.84
1330	3/4" x 12"	"	4.18
1400	Redwood Beveled - Clear Heart		
1410	1/2" x 6"	S.F.	5.86
1430	5/8" x 10"	"	5.97
1440	3/4" x 6"	"	6.08
1450	3/4" x 8"	"	6.76
1500	3/4" x 6" Board - Clear	"	3.55
1510	3/4" Tongue & Groove	"	5.53
1520	1" Rustic Beveled - 6" and 8"	"	6.12
1530	5/4" Rustic Beveled - 6" and 8"	"	6.05
1540	Add for Metal Corners	EA.	1.66
1550	Add for Mitering Corners	"	3.56
1600	Plywood Fir AC Smooth -One Side 1/4"	S.F.	2.09
1610	3/8"	"	2.27
1620	1/2"	"	2.31
1630	5/8"	"	2.46
1640	3/4"	"	2.69
1650	5/8" - Grooved and Rough Faced	"	2.79
1700	Cedar - Rough Sawn 3/8"	"	2.69
1710	5/8"	"	2.79
1720	3/4"	"	3.32
1730	5/8" - Grooved and Rough Faced	"	3.32
1740	Add for Wood Batten Strips, 1" x 2" - 4' O.C.	"	0.42
1750	Add for Splines	"	0.42
1800	Hardboard - Paneling and Lap Siding (Primed)		
1810	3/8" Rough Textured Paneling - 4' x 8'	S.F.	2.20
1820	7/16" Grooved - 4' x 8'	"	2.09
1830	7/16" Stucco Board Paneling	"	2.31
1840	7/16" x 8" Lap Siding	"	2.64
1860	Add for Pre-Finishing	"	0.21
1900	Shingles		
1910	Wood - 16" Red Cedar - 12" to Weather - #1	S.F.	4.29
1920	#2	"	3.44
1930	#3	"	3.22

		UNIT	COST
06999.20	**FINISH CARPENTRY, Cont'd...**		
1940	Red Cedar Hand Splits - #1, 24" - 1/2" to 3/4"	S.F.	3.65
1950	24" - 3/4" to 1 1/4"	"	4.06
1960	#2	"	3.65
1970	#3	"	3.44
2000	Add for Fire Retardant	"	0.51
2010	Add for 3/8" Backer Board	"	0.85
2020	Add for Metal Corners	EA.	1.64
2030	Add for Ridges, Hips and Corners	L.F.	4.45
2040	Add for 8" to Weather	PCT.	25.00
2100	Facia - Pine #2 1" x 8"	L.F.	2.04
2110	Cedar #3 1" x 8"	"	3.11
2120	Redwood - Clear 1" x 8"	"	4.41
2130	Plywood 5/8" - ACX 1" x 8"	"	2.09
2140	Facia - Pine #2 1" x 8"	S.F.	3.05
2200	Cedar #3 1" x 8"	"	4.56
2210	Redwood - Clear 1" x 8"	"	6.64
2220	Plywood 5/8" - ACX 1" x 8"	"	3.16
2300	FINISH WALLS (Interior) (15% Added for Waste)	"	3.44
2400	Boards, Cedar - #3 1" x 6" and 1" x 8"	"	3.44
2410	Knotty 1" x 6" and 1" x 8"	"	3.65
2420	D Grade 1" x 6" and 1" x 8"	"	4.54
2430	Aromatic 1" x 6" and 1" x 8"	"	4.73
2440	Redwood - Construction 1" x 6" and 1" x 8"	"	5.60
2450	Clear 1" x 6" and 1" x 8"	"	5.79
2460	Fir - Beaded 5/8" x 4"	"	4.93
2470	Pine - #2 1" x 6" and 1" x 8"	"	6.39
2500	Hardboard (Paneling) Tempered - 1/8"	"	4.29
2510	1/4"	"	3.38
2600	Plywood (Prefinished Paneling) 1/4" Birch - Natural	"	1.39
2610	3/4" Birch - Natural	"	1.57
2620	1/4" Birch - White	"	3.31
2630	1/4" Oak - Rotary Cut	"	4.36
2640	3/4"	"	4.09
2700	1/4" Oak - White	"	3.16
2710	1/4" Mahogany (Lauan)	"	4.55
2720	3/4"	"	4.36
2730	3/4" Mahogany (African)	"	2.30
2740	1/4" Walnut	"	3.42
2800	Gypsum Board (Paneling)	"	5.77
2810	Prefinished	"	6.28
2820	Red Oak Plastic		
06999.30	**MILLWORK & CUSTOM WOODWORK**		
1000	CUSTOM CABINET WORK (Red Oak or Birch)		
1010	Base Cabinets - Avg. 35" H x 24" D with Drawer	L.F.	207
1020	Sink Fronts	"	122
1030	Corner Cabinets	"	250
1040	Add per Drawer	"	53.00
1050	Upper Cabinets - Avg. 30" High x 12" Deep	"	158
2000	Utility Cabinets - Avg. 84" High x 24" Deep	"	380
2010	China or Corner Cabinets - 84" High	EA.	730
2020	Oven Cabinets - 84" High x 24" Deep	"	430

		UNIT	COST
06999.30	**MILLWORK & CUSTOM WOODWORK, Cont'd...**		
2030	Vanity Cabinets - Avg. 30" High x 21" Deep	EA.	250
2040	Deduct for Prefinishing wood	PCT.	12.36
2050	Base Cabinets - Avg. 35" H x 24" D with Drawer	EA.	198
2060	Sink Fronts	L.F.	123
2070	Corner Cabinets	"	280
2080	Add per Drawer	"	53.50
2090	Upper Cabinets - Avg. 30" High x 12" Deep	"	159
3000	Utility Cabinets - Avg. 84" High x 24" Deep	"	380
3010	China or Corner Cabinets - 84" High	"	740
3020	Oven Cabinets - 84" High x 24" Deep	"	420
3030	Vanity Cabinets - Avg. 30" High x 21" Deep	"	250
4000	COUNTER TOPS - 25"		
4010	Plastic Laminated with 4" Back Splash	L.F.	61.50
4020	Deduct for No Back Splash	"	11.06
4030	Granite - 1 1/4" - Artificial	"	123
4040	3/4" - Artificial	"	95.50
4050	Marble	"	122
4070	Wood Cutting Block	"	120
5000	Stainless Steel	"	187
5010	Polyester-Acrylic Solid Surface	"	196
5020	Quartz	"	106
5030	Plastic (Polymer)	"	75.25
6000	CUSTOM DOOR FRAMES - Including 2 Sides Trim		
6010	Birch	EA.	350
6020	Fir	"	195
6030	Poplar	"	198
6040	Oak	"	250
6050	Pine	"	240
6060	Walnut	"	460
8000	MOULDINGS AND TRIM		
8010	Apron 7/16" x 2"	L.F.	2.40
8020	Astragal 1 3/4" x 2 1/4"	"	5.83
8030	Base 7/16" x 2 3/4"	"	3.55
8050	Base Shoe 7/16" x 2 3/4"	"	2.91
8060	Batten Strip 5/8" x 1 5/8"	"	2.52
8070	Brick Mould 1 1/14" x 2"	"	2.93
9080	Casing 11/16" x 2 1/4"	"	3.21
9100	Chair Rail 5/8" x 1 3/4"	"	3.22
9110	Closet Rod 1 5/16"	"	3.66
9120	Corner Bead 1 1/8" x 1 1/8"	"	4.55
9130	Cove Moulding 3/4" x 3/4"	"	2.76
9140	Crown Moulding 9/16" x 3 5/8"	"	4.24
9160	Drip Cap	"	3.32
9170	Half Round 1/2" x 1"	"	3.91
9180	Hand Rail 1 5/8" x 1 3/4"	"	4.29
9190	Hook Strip 5/8" x 2 1/2"	"	2.91
9200	Picture Mould 3/4" x 1 1/2"	"	3.22
9210	Quarter Round 3/4" x 3/4"	"	2.26
9220	1/2" x 1/2"	"	1.74
9230	Sill 3/4" x 2 1/2"	"	3.90
9240	Stool 11/16 x 2 1/2"	"	4.66
9250	BIRCH		

		UNIT	COST
06999.30	**MILLWORK & CUSTOM WOODWORK, Cont'd...**		
9260	STAIRS (Treads, Risers, Skirt Boards)	L.F.	42.75
9270	SHELVING (12" Deep)	S.F.	45.50
9280	CUSTOM PANELING	"	30.75
9290	OAK		
9300	STAIRS (Treads, Risers, Skirt Boards)	L.F.	33.75
9310	SHELVING (12" Deep)	S.F.	40.75
9320	CUSTOM PANELING	"	22.99
9330	THRESHOLDS - 3/4" x 3 1/2"	L.F.	22.08
9340	PINE		
9350	STAIRS (Treads, Risers, Skirt Boards)	L.F.	25.25
9360	SHELVING (12" Deep)	"	28.00
9370	CUSTOM PANELING	S.F.	19.15
06999.40	**GLUE LAMINATE**		
0010	Arches 80' Span	S.F.	11.02
0020	Beams & Purlins 40' Span	"	7.57
0030	Deck Fir - 3" x 6"	"	9.37
0040	4" x 6"	"	10.42
0050	Deck Cedar - 3" x 6"	"	11.06
0060	4" x 6"	"	11.95
0070	Deck Pine - 3" x 6"	"	8.59
06999.50	**PREFABRICATED WOOD COMPONENTS**		
1000	WOOD TRUSSED RAFTERS		
1010	24" O.C. - 4 -12 Pitch		
2000	Span To 16' w/ Supports	EA.	147
2010	20'	"	151
2030	24'	"	180
2050	28'	"	191
2070	32'	"	250
3000	Add for 5-12 Pitch	PCT.	5.00
3010	Add for 6-12 Pitch	"	12.00
3020	Add for Scissor Truss	EA.	32.50
3030	Add for Gable End 24'	"	70.75
3040	Span To 16' w/ Supports	S.F.	4.67
3050	20'	"	3.71
3070	24'	"	3.53
3100	28'	"	3.49
3120	32'	"	3.99
5000	Add for 5-12 Pitch	PCT.	15.00
5010	Add for 6-12 Pitch	"	25.00
5020	Add for Scissor Truss	S.F.	25.50
5030	Add for Gable End 24'	"	0.46
6000	WOOD FLOOR TRUSS JOISTS (TJI)		
6010	Open Web - Wood or Metal - 24" O.C.		
6020	Up To 23' Span x 12" Single	L.F.	9.24
6030	To 24' Span x 15" Cord	"	9.49
6050	To 30' Span x 21"	"	10.29
6060	To 25' Span x 15" Double	"	9.06
6070	To 30' Span x 18" Cord	"	9.49
6080	To 36' Span x 21"	"	10.61
7000	Plywood Web - 24" O.C.		
7010	Up To 15' x 9 1/2"	L.F.	5.98

		UNIT	COST
06999.50	**PREFABRICATED WOOD COMPONENTS, Cont'd...**		
7030	To 21' x 14"	L.F.	6.68
7040	To 22' x 16"	"	7.48
8000	Open Web - Wood or Metal - 24" O.C.		
8010	Up To 23' Span x 12" Single	S.F.	4.61
8020	To 24' Span x 15" Cord	"	4.74
8030	To 27' Span x 18"	"	4.86
8040	To 30' Span x 21"	"	5.14
8050	To 25' Span x 15" Double	"	4.54
8060	To 30' Span x 18" Cord	"	4.61
8070	To 36' Span x 21"	"	5.21
8500	Plywood Web - 24" O.C.		
8510	Up To 15' x 9 1/2"	S.F.	2.97
8520	To 19' x 11 7/8"	"	3.00
8530	To 21' x 14"	"	3.34
8540	To 22' x 16"	"	3.71
06999.51	**LAMINATED VENEER STRUCTURAL BEAMS**		
0005	Parallel Strand Beams - Plywood - 24" O.C.		
0010	9 1/2" x 1 3/4"	L.F.	8.38
0020	11 7/8" x 1 3/4"	"	9.10
0030	14" x 1 3/4"	"	10.74
0040	16" x 1 3/4"	"	11.95
1000	Laminated Veneer Beams - Dimension Lumber- 24" O.C.		
1010	9" x 3 1/2"	L.F.	15.79
1020	12" x 3 1/2"	"	18.15
1040	9" x 5 1/2"	"	20.36
1050	12" x 5 1/2"	"	24.50
1070	18" x 5 1/2"	"	38.50
2000	Add for Architectural Grade	PCT.	25.00
2010	Add for 6 3/4"	"	40.00
3000	Laminated Strand Beams - Plywood - 24" O.C.		
3010	9 1/2" x 1 3/4"	S.F.	4.22
3020	11 7/8" x 1 3/4"	"	4.55
3040	16" x 1 3/4"	"	5.97
4000	Glue Lam Beam - Dimension Lumber - 24" O.C.		
4010	9" x 3 1/2"	S.F.	7.89
4020	12" x 3 1/2"	"	9.10
4040	9" x 5 1/2"	"	10.17
4050	12" x 5 1/2"	"	12.41
4070	18" x 5 1/2"	"	19.06
06999.60	**WOOD TREATMENTS**		
0010	PRESERVATIVES – PRESSURE TREATED		
0020	Dimensions	B.F.	0.23
0030	Timbers	"	0.37
0040	FIRE RETARDENTS		
0050	Dimensions & Timbers	B.F.	0.46
06999.70	**ROUGH HARDWARE**		
1000	NAILS	B.F.	0.01
2000	JOIST HANGERS	EA.	3.83
3000	BOLTS - 5/8" x 12"	"	4.50

Design & Construction Resources

TABLE OF CONTENTS   PAGE

07100.10 - WATERPROOFING — 07-2
07160.10 - BITUMINOUS DAMPPROOFING — 07-2
07190.10 - VAPOR BARRIERS — 07-2
07210.10 - BATT INSULATION — 07-2
07210.20 - BOARD INSULATION — 07-3
07210.60 - LOOSE FILL INSULATION — 07-3
07210.70 - SPRAYED INSULATION — 07-3
07310.10 - ASPHALT SHINGLES — 07-3
07310.50 - METAL SHINGLES — 07-3
07310.60 - SLATE SHINGLES — 07-4
07310.70 - WOOD SHINGLES — 07-4
07310.80 - WOOD SHAKES — 07-4
07310.90 - CLAY TILE ROOF (assorted colors) — 07-4
07460.10 - METAL SIDING PANELS — 07-4
07460.50 - PLASTIC SIDING — 07-5
07460.60 - PLYWOOD SIDING — 07-5
07460.80 - WOOD SIDING — 07-5
07510.10 - BUILT-UP ASPHALT ROOFING — 07-6
07530.10 - SINGLE-PLY ROOFING — 07-6
07610.10 - METAL ROOFING — 07-6
07620.10 - FLASHING AND TRIM — 07-6
07620.20 - GUTTERS AND DOWNSPOUTS — 07-7
07810.10 - PLASTIC SKYLIGHTS — 07-8
07920.10 - CAULKING — 07-8
QUICK ESTIMATING — **07A-9**

DIVISION # 07 THERMAL AND MOISTURE

		UNIT	LABOR	MAT.	TOTAL
07100.10	**WATERPROOFING**				
0100	Membrane waterproofing, elastomeric				
1020	Butyl				
1040	1/32" thick	S.F.	1.50	1.05	2.55
1060	1/16" thick	"	1.56	1.37	2.93
1140	Neoprene				
1160	1/32" thick	S.F.	1.50	1.71	3.21
1180	1/16" thick	"	1.56	2.80	4.36
1260	Plastic vapor barrier (polyethylene)				
1280	4 mil	S.F.	0.15	0.02	0.17
1300	6 mil	"	0.15	0.03	0.18
1320	10 mil	"	0.18	0.09	0.27
1400	Bituminous membrane waterproofing, asphalt felt, 15 lb.				
1440	One ply	S.F.	0.93	0.62	1.55
1460	Two ply	"	1.13	0.73	1.86
1480	Three ply	"	1.34	0.91	2.25
07160.10	**BITUMINOUS DAMPPROOFING**				
0100	Building paper, asphalt felt				
0120	15 lb	S.F.	1.50	0.15	1.65
0140	30 lb	"	1.56	0.28	1.84
1000	Asphalt dampproofing, troweled, cold, primer plus				
1020	1 coat	S.F.	1.25	0.68	1.93
1040	2 coats	"	1.87	1.43	3.30
1060	3 coats	"	2.34	2.04	4.38
1200	Fibrous asphalt dampproofing, hot troweled, primer plus				
1220	1 coat	S.F.	1.50	0.68	2.18
1240	2 coats	"	2.08	1.43	3.51
1260	3 coats	"	2.68	2.04	4.72
07190.10	**VAPOR BARRIERS**				
0980	Vapor barrier, polyethylene				
1000	2 mil	S.F.	0.18	0.01	0.19
1010	6 mil	"	0.18	0.04	0.22
1020	8 mil	"	0.20	0.05	0.25
1040	10 mil	"	0.20	0.06	0.26
07210.10	**BATT INSULATION**				
0980	Ceiling, fiberglass, unfaced				
1000	3-1/2" thick, R11	S.F.	0.44	0.38	0.82
1020	6" thick, R19	"	0.50	0.50	1.00
1030	9" thick, R30	"	0.57	1.00	1.57
1035	Suspended ceiling, unfaced				
1040	3-1/2" thick, R11	S.F.	0.41	0.38	0.79
1060	6" thick, R19	"	0.46	0.50	0.96
1070	9" thick, R30	"	0.53	1.00	1.53
1075	Crawl space, unfaced				
1080	3-1/2" thick, R11	S.F.	0.57	0.38	0.95
1100	6" thick, R19	"	0.62	0.50	1.12
1120	9" thick, R30	"	0.68	1.00	1.68
2000	Wall, fiberglass				
2010	Paper backed				
2020	2" thick, R7	S.F.	0.39	0.25	0.64
2040	3" thick, R8	"	0.41	0.27	0.68
2060	4" thick, R11	"	0.44	0.45	0.89

		UNIT	LABOR	MAT.	TOTAL
07210.10	**BATT INSULATION, Cont'd...**				
2080	6" thick, R19	S.F.	0.46	0.67	1.13
2090	Foil backed, 1 side				
2100	2" thick, R7	S.F.	0.39	0.58	0.97
2120	3" thick, R11	"	0.41	0.61	1.02
2140	4" thick, R14	"	0.44	0.64	1.08
2160	6" thick, R21	"	0.46	0.67	1.13
07210.20	**BOARD INSULATION**				
2200	Perlite board, roof				
2220	1.00" thick, R2.78	S.F.	0.31	0.77	1.08
2240	1.50" thick, R4.17	"	0.32	1.19	1.51
2580	Rigid urethane				
2600	1" thick, R6.67	S.F.	0.31	1.03	1.34
2640	1.50" thick, R11.11	"	0.32	1.40	1.72
2780	Polystyrene				
2800	1.0" thick, R4.17	S.F.	0.31	0.41	0.72
2820	1.5" thick, R6.26	"	0.32	0.63	0.95
07210.60	**LOOSE FILL INSULATION**				
1000	Blown-in type				
1010	Fiberglass				
1020	5" thick, R11	S.F.	0.31	0.36	0.67
1040	6" thick, R13	"	0.37	0.41	0.78
1060	9" thick, R19	"	0.53	0.50	1.03
07210.70	**SPRAYED INSULATION**				
1000	Foam, sprayed on				
1010	Polystyrene				
1020	1" thick, R4	S.F.	0.37	0.44	0.81
1040	2" thick, R8	"	0.50	0.85	1.35
1050	Urethane				
1060	1" thick, R4	S.F.	0.37	0.41	0.78
1080	2" thick, R8	"	0.50	0.79	1.29
07310.10	**ASPHALT SHINGLES**				
1000	Standard asphalt shingles, strip shingles				
1020	210 lb/square	SQ.	46.25	68.00	114
1040	235 lb/square	"	52.00	72.00	124
1060	240 lb/square	"	58.00	75.00	133
1080	260 lb/square	"	66.00	110	176
1100	300 lb/square	"	77.00	120	197
1120	385 lb/square	"	93.00	160	253
5980	Roll roofing, mineral surface				
6000	90 lb	SQ.	33.00	38.50	71.50
6020	110 lb	"	38.75	64.00	103
6040	140 lb	"	46.25	65.00	111
07310.50	**METAL SHINGLES**				
0980	Aluminum, .020" thick				
1000	Plain	SQ.	93.00	260	353
1020	Colors	"	93.00	280	373
1960	Steel, galvanized				
1980	26 ga.				
2000	Plain	SQ.	93.00	260	353
2020	Colors	"	93.00	340	433
2030	24 ga.				

		UNIT	LABOR	MAT.	TOTAL
07310.50	**METAL SHINGLES, Cont'd...**				
2040	Plain	SQ.	93.00	280	373
2060	Colors	"	93.00	350	443
07310.60	**SLATE SHINGLES**				
0960	Slate shingles				
0980	Pennsylvania				
1000	Ribbon	SQ.	230	630	860
1020	Clear	"	230	810	1,040
1030	Vermont				
1040	Black	SQ.	230	680	910
1060	Gray	"	230	750	980
1070	Green	"	230	760	990
1080	Red	"	230	1,380	1,610
07310.70	**WOOD SHINGLES**				
1000	Wood shingles, on roofs				
1010	White cedar, #1 shingles				
1020	4" exposure	SQ.	150	230	380
1040	5" exposure	"	120	210	330
1050	#2 shingles				
1060	4" exposure	SQ.	150	160	310
1080	5" exposure	"	120	140	260
1090	Resquared and rebutted				
1100	4" exposure	SQ.	150	210	360
1120	5" exposure	"	120	170	290
1140	On walls				
1150	White cedar, #1 shingles				
1160	4" exposure	SQ.	230	230	460
1180	5" exposure	"	190	210	400
1200	6" exposure	"	150	170	320
1210	#2 shingles				
1220	4" exposure	SQ.	230	160	390
1240	5" exposure	"	190	140	330
1260	6" exposure	"	150	120	270
1300	Add for fire retarding	"			110
07310.80	**WOOD SHAKES**				
2010	Shakes, hand split, 24" red cedar, on roofs				
2020	5" exposure	SQ.	230	240	470
2040	7" exposure	"	190	230	420
2060	9" exposure	"	150	210	360
2080	On walls				
2100	6" exposure	SQ.	230	220	450
2120	8" exposure	"	190	210	400
2140	10" exposure	"	150	200	350
3000	Add for fire retarding	"			75.00
07310.90	**CLAY TILE ROOF (assorted colors)**				
1010	Spanish tile, terracota	SQ.	120	390	510
1020	Mission tile, terracota	"	150	770	920
1030	French tile, rustic blue, green patina	"	130	880	1,010
07460.10	**METAL SIDING PANELS**				
1000	Aluminum siding panels				
1020	Corrugated				
1030	Plain finish				

DIVISION # 07 THERMAL AND MOISTURE

		UNIT	LABOR	MAT.	TOTAL
07460.10	**METAL SIDING PANELS, Cont'd...**				
1040	.024"	S.F.	2.14	1.54	3.68
1060	.032"	"	2.14	1.81	3.95
1070	Painted finish				
1080	.024"	S.F.	2.14	1.92	4.06
1100	.032"	"	2.14	2.20	4.34
2000	Steel siding panels				
2040	Corrugated				
2080	22 ga.	S.F.	3.58	1.95	5.53
2100	24 ga.	"	3.58	1.79	5.37
07460.50	**PLASTIC SIDING**				
1000	Horizontal vinyl siding, solid				
1010	8" wide				
1020	Standard	S.F.	1.85	1.23	3.08
1040	Insulated	"	1.85	1.49	3.34
1050	10" wide				
1060	Standard	S.F.	1.72	1.27	2.99
1080	Insulated	"	1.72	1.52	3.24
8500	Vinyl moldings for doors and windows	L.F.	1.93	0.79	2.72
07460.60	**PLYWOOD SIDING**				
1000	Rough sawn cedar, 3/8" thick	S.F.	1.60	1.79	3.39
1020	Fir, 3/8" thick	"	1.60	0.99	2.59
1980	Texture 1-11, 5/8" thick				
2000	Cedar	S.F.	1.72	2.42	4.14
2020	Fir	"	1.72	1.69	3.41
2040	Redwood	"	1.65	2.59	4.24
2060	Southern Yellow Pine	"	1.72	1.37	3.09
07460.80	**WOOD SIDING**				
1000	Beveled siding, cedar				
1010	A grade				
1040	1/2 x 8	S.F.	1.93	3.68	5.61
1060	3/4 x 10	"	1.60	4.74	6.34
1070	Clear				
1100	1/2 x 8	S.F.	1.93	4.10	6.03
1120	3/4 x 10	"	1.60	5.50	7.10
1130	B grade				
1160	1/2 x 8	S.F.	19.25	4.37	23.62
1180	3/4 x 10	"	1.60	4.12	5.72
2000	Board and batten				
2010	Cedar				
2020	1x6	S.F.	2.41	4.26	6.67
2040	1x8	"	1.93	3.88	5.81
2060	1x10	"	1.72	3.50	5.22
2080	1x12	"	1.55	3.14	4.69
2090	Pine				
2100	1x6	S.F.	2.41	1.29	3.70
2120	1x8	"	1.93	1.26	3.19
2140	1x10	"	1.72	1.21	2.93
2160	1x12	"	1.55	1.12	2.67
3000	Tongue and groove				
3010	Cedar				
3020	1x4	S.F.	2.68	4.80	7.48

DIVISION # 07 THERMAL AND MOISTURE

		UNIT	LABOR	MAT.	TOTAL
07460.80	**WOOD SIDING, Cont'd...**				
3040	1x6	S.F.	2.54	4.63	7.17
3060	1x8	"	2.41	4.33	6.74
3080	1x10	"	2.29	4.26	6.55
3090	Pine				
3100	1x4	S.F.	2.68	1.44	4.12
3120	1x6	"	2.54	1.36	3.90
3140	1x8	"	2.41	1.27	3.68
3160	1x10	"	2.29	1.21	3.50
07510.10	**BUILT-UP ASPHALT ROOFING**				
0980	Built-up roofing, asphalt felt, including gravel				
1000	2 ply	SQ.	120	56.00	176
1500	3 ply	"	150	76.00	226
2000	4 ply	"	190	110	300
2195	Cant strip, 4" x 4"				
2200	Treated wood	L.F.	1.32	1.92	3.24
2260	Foamglass	"	1.15	1.03	2.18
8000	New gravel for built-up roofing, 400 lb/sq	SQ.	93.00	32.00	125
07530.10	**SINGLE-PLY ROOFING**				
2000	Elastic sheet roofing				
2060	Neoprene, 1/16" thick	S.F.	0.57	2.69	3.26
2115	PVC				
2120	45 mil	S.F.	0.57	2.03	2.60
2200	Flashing				
2220	Pipe flashing, 90 mil thick				
2260	1" pipe	EA.	11.50	28.50	40.00
2360	Neoprene flashing, 60 mil thick strip				
2380	6" wide	L.F.	3.86	1.73	5.59
2390	12" wide	"	5.79	3.42	9.21
07610.10	**METAL ROOFING**				
1000	Sheet metal roofing, copper, 16 oz, batten seam	SQ.	310	1,000	1,310
1020	Standing seam	"	290	970	1,260
2000	Aluminum roofing, natural finish				
2005	Corrugated, on steel frame				
2010	.0175" thick	SQ.	130	120	250
2040	.0215" thick	"	130	160	290
2060	.024" thick	"	130	190	320
2080	.032" thick	"	130	240	370
2100	V-beam, on steel frame				
2120	.032" thick	SQ.	130	230	360
2130	.040" thick	"	130	240	370
2140	.050" thick	"	130	310	440
2200	Ridge cap				
2220	.019" thick	L.F.	1.54	3.85	5.39
2500	Corrugated galvanized steel roofing, on steel frame				
2520	28 ga.	SQ.	130	120	250
2540	26 ga.	"	130	140	270
2550	24 ga.	"	130	160	290
2560	22 ga.	"	130	170	300

		UNIT	LABOR	MAT.	TOTAL
07620.10	**FLASHING AND TRIM**				
0050	Counter flashing				
0060	Aluminum, .032"	S.F.	4.63	1.44	6.07
0100	Stainless steel, .015"	"	4.63	4.62	9.25
0105	Copper				
0110	16 oz.	S.F.	4.63	7.15	11.78
0112	20 oz.	"	4.63	8.49	13.12
0114	24 oz.	"	4.63	10.25	14.88
0116	32 oz.	"	4.63	12.50	17.13
0118	Valley flashing				
0120	Aluminum, .032"	S.F.	2.89	1.52	4.41
0130	Stainless steel, .015	"	2.89	4.81	7.70
0135	Copper				
0140	16 oz.	S.F.	2.89	6.78	9.67
0160	20 oz.	"	3.86	8.49	12.35
0180	24 oz.	"	2.89	10.25	13.14
0200	32 oz.	"	2.89	12.50	15.39
0380	Base flashing				
0400	Aluminum, .040"	S.F.	3.86	2.60	6.46
0410	Stainless steel, .018"	"	3.86	5.78	9.64
0415	Copper				
0420	16 oz.	S.F.	3.86	6.78	10.64
0422	20 oz.	"	2.89	8.49	11.38
0424	24 oz.	"	3.86	10.25	14.11
0426	32 oz.	"	3.86	12.50	16.36
07620.20	**GUTTERS AND DOWNSPOUTS**				
1500	Copper gutter and downspout				
1520	Downspouts, 16 oz. copper				
1530	Round				
1540	3" dia.	L.F.	3.09	10.50	13.59
1550	4" dia.	"	3.09	13.00	16.09
1800	Gutters, 16 oz. copper				
1810	Half round				
1820	4" wide	L.F.	4.63	9.38	14.01
1840	5" wide	"	5.15	11.50	16.65
1860	Type K				
1880	4" wide	L.F.	4.63	10.50	15.13
1890	5" wide	"	5.15	11.00	16.15
3000	Aluminum gutter and downspout				
3005	Downspouts				
3010	2" x 3"	L.F.	3.09	1.04	4.13
3030	3" x 4"	"	3.31	1.43	4.74
3035	4" x 5"	"	3.56	1.59	5.15
3038	Round				
3040	3" dia.	L.F.	3.09	1.76	4.85
3050	4" dia.	"	3.31	2.25	5.56
3240	Gutters, stock units				
3260	4" wide	L.F.	4.88	1.70	6.58
3270	5" wide	"	5.15	2.03	7.18
4101	Galvanized steel gutter and downspout				
4111	Downspouts, round corrugated				
4121	3" dia.	L.F.	3.09	1.76	4.85

		UNIT	LABOR	MAT.	TOTAL
07620.20	**GUTTERS AND DOWNSPOUTS, Cont'd...**				
4131	4" dia.	L.F.	3.09	2.36	5.45
4141	5" dia.	"	3.31	3.52	6.83
4151	6" dia.	"	3.31	4.67	7.98
4161	Rectangular				
4171	2" x 3"	L.F.	3.09	1.59	4.68
4191	3" x 4"	"	2.89	2.28	5.17
4201	4" x 4"	"	2.89	2.86	5.75
4300	Gutters, stock units				
4310	5" wide				
4320	Plain	L.F.	5.15	1.61	6.76
4330	Painted	"	5.15	1.76	6.91
4335	6" wide				
4340	Plain	L.F.	5.45	2.25	7.70
4360	Painted	"	5.45	2.53	7.98
07810.10	**PLASTIC SKYLIGHTS**				
1030	Single thickness, not including mounting curb				
1040	2' x 4'	EA.	58.00	360	418
1050	4' x 4'	"	77.00	490	567
1060	5' x 5'	"	120	650	770
1070	6' x 8'	"	150	1,390	1,540
07920.10	**CAULKING**				
0100	Caulk exterior, two component				
0120	1/4 x 1/2	L.F.	2.41	0.39	2.80
0140	3/8 x 1/2	"	2.68	0.60	3.28
0160	1/2 x 1/2	"	3.01	0.82	3.83
0220	Caulk interior, single component				
0240	1/4 x 1/2	L.F.	2.29	0.26	2.55
0260	3/8 x 1/2	"	2.54	0.37	2.91
0280	1/2 x 1/2	"	2.84	0.49	3.33

		UNIT	COST
07999.10	**WATERPROOFING**		
0010	1-Ply Membrane Felt 15#	S.F.	1.63
0020	2-Ply Membrane Felt 15#	"	2.49
0030	Hydrolithic	"	2.68
0040	Elastomeric Rubberized Asphalt with Poly Sheet	"	2.49
0060	Metallic Oxide 3-Coat	"	4.24
0070	Vinyl Plastic	"	2.96
0080	Bentonite 3/8" - Trowel	"	1.98
0090	5/8" - Panels	"	2.24
07999.20	**DAMPPROOFING**		
0010	Asphalt Trowel Mastic 1/16"	S.F.	0.99
0020	1/8"	"	1.20
0030	Spray Liquid 1-Coat	"	0.62
0040	2-Coat	"	0.87
0050	Brush Liquid 2-Coat	"	0.97
1000	Hot Mop 1-Coat and Primer	"	1.63
1010	1 Fibrous Asphalt	"	1.78
1020	Cementitious Per Coat 1/2" Coat	"	0.80
1030	Silicone 1-Coat	"	0.58
1040	2-Coat	"	0.90
1050	Add for Scaffold and Lift Operations	"	0.50
07999.30	**BUILDING INSULATION**		
1000	FLEXIBLE		
1010	Fiberglass 2 1/4" R 7.40	S.F.	0.74
1020	3 1/2" R 11.00	"	0.85
1040	6" R 19.00	"	1.06
1060	12" R 38.00	"	1.49
1100	Add for Ceiling Work	"	0.09
1110	Add for Paper Faced	"	0.11
1120	Add for Polystyrene Barrier (2m)	"	0.11
1130	Add for Scaffold Work	"	0.61
1200	RIGID		
1300	Fiberglass 1" R 4.35 3# Density	S.F.	1.12
1310	1 1/2" R 6.52 3# Density	"	1.38
1320	2" R 8.70 3# Density	"	1.71
1400	Styrene, Molded 1" R 4.3.5	"	8.54
1410	1 1/2" R 6.52	"	0.96
1420	2" R 7.69	"	1.28
1430	2" T&G R 7.69	"	1.65
1500	Styrene, Extruded 1" R 5.40	"	1.18
1510	1 1/2" R 6.52	"	1.38
1520	2" R 7.69	"	1.49
1530	2" T&G R 7.69	"	1.65
1600	Perlite 1" R 2.78	"	1.18
1610	2" R 5.56	"	1.87
1700	Urethane 1" R 6.67	"	1.38
1710	2" R 13.34	"	1.76
1720	Add for Glued Applications	"	0.09
1800	LOOSE 1" Styrene R 3.8	"	0.82
1810	1" Fiberglass R 2.2	"	0.78
1820	1" Rock Wool R 2.9	"	0.80
1830	1" Cellulose R 3.7	"	0.76

		UNIT	COST
07999.30	**BUILDING INSULATION, Cont'd...**		
1900	FOAMED Urethane Per Inch	S.F.	2.01
2000	SPRAYED Cellulose Per Inch	"	1.46
2100	Polystyrene	"	1.74
2200	Urethane	"	2.47
2300	ALUMINUM PAPER	"	0.66
2400	VINYL FACED FIBERGLASS	"	1.98
07999.80	**SHINGLE ROOFING**		
1000	ASPHALT SHINGLES 235# Seal Down	S.F.	1.67
1010	300# Laminated	"	1.01
1020	325# Fire Resistant	"	2.18
1030	Timberline	"	2.13
1040	340# Tab Lock	"	2.81
2000	ASPHALT ROLL 90#	"	0.94
3000	FIBERGLASS SHINGLES 215#	"	1.56
3010	250#	"	1.91
4000	RED CEDAR 16" x 5" to Weather #1 Grade	"	3.87
4010	#2 Grade	"	3.44
4020	#3 Grade	"	2.46
4030	24" x 10" to Weather - Hand Splits - 1/2" x 3/4" #	"	25.27
4040	3/4" x 1 1/4" #	"	25.52
4050	Add for Fire Retardant	"	0.65
5000	METAL Aluminum 020 mil	"	4.23
6000	Anodized 020 Mil	"	4.90
7000	Steel, Enameled Colored	"	6.30
8000	Galvanized Colored	"	5.45
9000	Add to Above for 15# Felt Underlayment		
9010	Add for Base Starter	S.F.	0.77
9020	Add for Ice & Water Starter	"	1.25
9030	Add for Boston Ridge	"	2.85
9040	Add for Removal & Haul Away	"	1.09
9050	Add for Pitches Over 5-12 each Pitch Increase	"	0.06
9060	Add for Chimneys, Skylights and Bay Windows	EA.	82.00
07999.90	**BUILT-UP ROOFING**		
1000	MEMBRANE		
1010	3-Ply Asphalt & Gravel - R 10.0	S.F.	670
1020	R 16.6	"	690
1030	4-Ply - R 10.0	"	690
1040	R 16.6	"	710
1050	5-Ply - R 10.0	"	730
1060	R 16.6	"	760
2000	Add for Sheet Rock over Steel Deck - 5/8"	"	145
2010	Add for Fiberglass Insulation	"	48.75
2020	Add for Pitch and Gravel	"	75.00
2030	Add for Sloped Roofs	"	53.00
2040	Add for Thermal Barrier	"	41.00
2050	Add for Upside Down Roofing System	"	145
3000	SINGLE-PLY - 60M Butylene Roofing & Gravel Ballast - R 10.0	"	530
3010	Mech. Fastened - R 10.0	"	513
3020	PVC & EPDM Roofing & Gravel Ballast - R 10.0	"	493
3030	Mech. Fastened - R 10.0	"	513
3040	Blocking and Cants Not Included		

TABLE OF CONTENTS   PAGE

08110.10 - METAL DOORS	08-2
08110.40 - METAL DOOR FRAMES	08-2
08210.10 - WOOD DOORS	08-3
08210.90 - WOOD FRAMES	08-5
08300.10 - SPECIAL DOORS	08-6
08410.10 - STOREFRONTS	08-8
08510.10 - STEEL WINDOWS	08-8
08520.10 - ALUMINUM WINDOWS	08-8
08600.10 - WOOD WINDOWS	08-9
08710.10 - HINGES	08-10
08710.20 - LOCKSETS	08-10
08710.30 - CLOSERS	08-11
08710.40 - DOOR TRIM	08-11
08710.60 - WEATHERSTRIPPING	08-11
08810.10 - GLAZING	08-12
08910.10 - GLAZED CURTAIN WALLS	08-13
QUICK ESTIMATING	**08A-15**

		UNIT	LABOR	MAT.	TOTAL
08110.10	**METAL DOORS**				
1000	Flush hollow metal, std. duty, 20 ga., 1-3/8" thick				
1020	2-6 x 6-8	EA.	54.00	220	274
1080	3-0 x 6-8	"	54.00	280	334
1090	1-3/4" thick				
1100	2-6 x 6-8	EA.	54.00	260	314
1150	3-0 x 6-8	"	54.00	300	354
1200	2-6 x 7-0	"	54.00	290	344
1240	3-0 x 7-0	"	54.00	310	364
2110	Heavy duty, 20 ga., unrated, 1-3/4"				
2130	2-8 x 6-8	EA.	54.00	280	334
2135	3-0 x 6-8	"	54.00	300	354
2140	2-8 x 7-0	"	54.00	300	354
2150	3-0 x 7-0	"	54.00	320	374
2200	18 ga., 1-3/4", unrated door				
2210	2-0 x 7-0	EA.	54.00	310	364
2235	2-6 x 7-0	"	54.00	330	384
2260	3-0 x 7-0	"	54.00	360	414
2270	3-4 x 7-0	"	54.00	420	474
2310	2", unrated door				
2320	2-0 x 7-0	EA.	60.00	320	380
2340	2-6 x 7-0	"	60.00	350	410
2360	3-0 x 7-0	"	60.00	380	440
2370	3-4 x 7-0	"	60.00	460	520
2400	Galvanized metal door				
2410	3-0 x 7-0	EA.	60.00	480	540
2450	For lead lining in doors	"			970
2460	For sound attenuation	"			45.00
4280	Vision glass				
4300	8" x 8"	EA.	60.00	100	160
4320	8" x 48"	"	60.00	150	210
4340	Fixed metal louver	"	48.25	160	208
4350	For fire rating, add				
4370	3 hr door	EA.			180
4380	1-1/2 hr door	"			76.00
4400	3/4 hr door	"			37.75
4430	1' extra height, add to material, 20%				
4440	1'6" extra height, add to material, 60%				
4470	For dutch doors with shelf, add to material, 100%				
5000	Stainless steel, general application				
5010	3'x7'	EA.	540	7,150	7,690
5020	5'X7'	"	720	12,650	13,370
6000	Heavy impact, s-core, 18 ga., stainless				
6010	3'x7'	EA.	540	880	1,420
6020	5'X7'	"	720	1,650	2,370
08110.40	**METAL DOOR FRAMES**				
1000	Hollow metal, stock, 18 ga., 4-3/4" x 1-3/4"				
1020	2-0 x 7-0	EA.	60.00	130	190
1060	2-6 x 7-0	"	60.00	150	210
1100	3-0 x 7-0	"	60.00	160	220
1120	4-0 x 7-0	"	80.00	160	240
1140	5-0 x 7-0	"	80.00	170	250

DIVISION # 08 DOORS AND WINDOWS

		UNIT	LABOR	MAT.	TOTAL
08110.40	**METAL DOOR FRAMES, Cont'd...**				
1160	6-0 x 7-0	EA.	80.00	180	260
1500	16 ga., 6-3/4" x 1-3/4"				
1520	2-0 x 7-0	EA.	67.00	140	207
1535	2-6 x 7-0	"	67.00	150	217
1550	3-0 x 7-0	"	67.00	160	227
1560	4-0 x 7-0	"	90.00	170	260
1580	6-0 x 7-0	"	90.00	180	270
08210.10	**WOOD DOORS**				
0980	Solid core, 1-3/8" thick				
1000	Birch faced				
1020	2-4 x 7-0	EA.	60.00	130	190
1060	3-0 x 7-0	"	60.00	140	200
1070	3-4 x 7-0	"	60.00	280	340
1080	2-4 x 6-8	"	60.00	130	190
1090	2-6 x 6-8	"	60.00	130	190
1100	3-0 x 6-8	"	60.00	140	200
1120	Lauan faced				
1140	2-4 x 6-8	EA.	60.00	140	200
1180	3-0 x 6-8	"	60.00	150	210
1200	3-4 x 6-8	"	60.00	150	210
1300	Tempered hardboard faced				
1320	2-4 x 7-0	EA.	60.00	140	200
1360	3-0 x 7-0	"	60.00	160	220
1380	3-4 x 7-0	"	60.00	180	240
1420	Hollow core, 1-3/8" thick				
1440	Birch faced				
1460	2-4 x 7-0	EA.	60.00	130	190
1500	3-0 x 7-0	"	60.00	130	190
1520	3-4 x 7-0	"	60.00	140	200
1600	Lauan faced				
1620	2-4 x 6-8	EA.	60.00	53.00	113
1630	2-6 x 6-8	"	60.00	57.00	117
1660	3-0 x 6-8	"	60.00	75.00	135
1680	3-4 x 6-8	"	60.00	83.00	143
1740	Tempered hardboard faced				
1770	2-6 x 7-0	EA.	60.00	64.00	124
1800	3-0 x 7-0	"	60.00	75.00	135
1820	3-4 x 7-0	"	60.00	81.00	141
1900	Solid core, 1-3/4" thick				
1920	Birch faced				
1940	2-4 x 7-0	EA.	60.00	230	290
1950	2-6 x 7-0	"	60.00	240	300
1970	3-0 x 7-0	"	60.00	230	290
1980	3-4 x 7-0	"	60.00	240	300
2000	Lauan faced				
2020	2-4 x 7-0	EA.	60.00	160	220
2030	2-6 x 7-0	"	60.00	180	240
2060	3-4 x 7-0	"	60.00	200	260
2080	3-0 x 7-0	"	60.00	220	280
2140	Tempered hardboard faced				
2170	2-6 x 7-0	EA.	60.00	180	240

		UNIT	LABOR	MAT.	TOTAL
08210.10	**WOOD DOORS, Cont'd...**				
2190	3-0 x 7-0	EA.	60.00	210	270
2200	3-4 x 7-0	"	60.00	220	280
2250	Hollow core, 1-3/4" thick				
2270	Birch faced				
2295	2-6 x 7-0	EA.	60.00	130	190
2320	3-0 x 7-0	"	60.00	130	190
2340	3-4 x 7-0	"	60.00	140	200
2400	Lauan faced				
2430	2-6 x 6-8	EA.	60.00	83.00	143
2460	3-0 x 6-8	"	60.00	79.00	139
2480	3-4 x 6-8	"	60.00	81.00	141
2520	Tempered hardboard				
2550	2-6 x 7-0	EA.	60.00	72.00	132
2580	3-0 x 7-0	"	60.00	79.00	139
2600	3-4 x 7-0	"	60.00	82.00	142
2620	Add-on, louver	"	48.25	27.00	75.25
2640	Glass	"	48.25	83.00	131
2700	Exterior doors, 3-0 x 7-0 x 2-1/2", solid core				
2710	Carved				
2720	One face	EA.	120	1,160	1,280
2740	Two faces	"	120	1,590	1,710
3000	Closet doors, 1-3/4" thick				
3001	Bi-fold or bi-passing, includes frame and trim				
3020	Paneled				
3040	4-0 x 6-8	EA.	80.00	360	440
3060	6-0 x 6-8	"	80.00	380	460
3070	Louvered				
3080	4-0 x 6-8	EA.	80.00	260	340
3100	6-0 x 6-8	"	80.00	300	380
3130	Flush				
3140	4-0 x 6-8	EA.	80.00	230	310
3160	6-0 x 6-8	"	80.00	260	340
3170	Primed				
3180	4-0 x 6-8	EA.	80.00	260	340
3200	6-0 x 6-8	"	80.00	290	370
3210	French Door, Dual-Tempered, Clear-Glass, 6'-8'				
3220	24"x80"x1-3/4"	EA.	97.00	370	467
3230	with Low-E glass	"	97.00	500	597
3240	30"x80"x1-3/4"	"	97.00	420	517
3250	with Low-E glass	"	97.00	500	597
3260	42"x80"x1-3/4"	"	97.00	550	647
3270	with Low-E glass	"	97.00	660	757
3280	24"x96"x1-3/4"	"	97.00	440	537
3290	with Low-E glass	"	97.00	550	647
3310	32"x96"x1-3/4"	"	97.00	490	587
3320	with Low-E glass	"	97.00	610	707
3330	French door, 10-lite, 1-3/4' thick, 6'-8" high				
3340	24" wide	EA.	97.00	390	487
3350	30" wide	"	97.00	480	577
3360	36" wide	"	97.00	490	587
3370	96" high, 24" wide	"	97.00	500	597
3380	30" wide	"	97.00	600	697

		UNIT	LABOR	MAT.	TOTAL
08210.10	**WOOD DOORS, Cont'd...**				
3390	36" wide	EA.	97.00	780	877
3400	French door, 1-lite, 1-3/4' thick, 6'-8" high				
3410	48" wide	EA.	97.00	2,840	2,937
3420	56" wide	"	97.00	1,650	1,747
3430	60" wide	"	97.00	1,930	2,027
3440	72" wide	"	97.00	1,810	1,907
3500	Fiberglass, single door, 6'-8" high				
3510	2-4" wide	EA.	60.00	290	350
3520	2-6" wide	"	60.00	430	490
3530	2-8" wide	"	60.00	430	490
3540	3-0" wide	"	60.00	570	630
3550	8'-0" high				
3560	2-4" wide	EA.	60.00	570	630
3570	2-6" wide	"	60.00	720	780
3580	2-8" wide	"	60.00	860	920
3590	3-0" wide	"	60.00	860	920
3600	Fiberglass, double door, 6'-8" high				
3610	56" wide	EA.	97.00	1,000	1,097
3620	60" wide	"	97.00	1,150	1,247
3630	64" wide	"	97.00	1,190	1,287
3640	72" wide	"	97.00	1,360	1,457
3650	8'-0" high				
3700	56" wide	EA.	97.00	260	357
3710	60" wide	"	97.00	270	367
3720	64" wide	"	97.00	290	387
3730	72" wide	"	97.00	310	407
3740	Metal clad, double door, 6'-8" high				
3750	56" wide	EA.	97.00	280	377
3760	60" wide	"	97.00	290	387
3770	64" wide	"	97.00	300	397
3780	72" wide	"	97.00	310	407
3790	8'-0" high				
3800	56" wide	EA.	97.00	470	567
3810	60" wide	"	97.00	500	597
3820	64" wide	"	97.00	550	647
3830	72" wide	"	97.00	600	697
3840	For outswinging doors, add min.	"			66.00
3850	Maximum	"			140
08210.90	**WOOD FRAMES**				
0080	Frame, interior, pine				
0100	2-6 x 6-8	EA.	69.00	32.25	101
0160	3-0 x 6-8	"	69.00	35.75	105
0180	5-0 x 6-8	"	69.00	37.50	107
0200	6-0 x 6-8	"	69.00	39.50	109
0220	2-6 x 7-0	"	69.00	36.75	106
0260	3-0 x 7-0	"	69.00	43.25	112
0280	5-0 x 7-0	"	97.00	47.50	145
0300	6-0 x 7-0	"	97.00	49.00	146
1000	Exterior, custom, with threshold, including trim				
1040	Walnut				
1060	3-0 x 7-0	EA.	120	310	430

		UNIT	LABOR	MAT.	TOTAL
08210.90	**WOOD FRAMES, Cont'd...**				
1080	6-0 x 7-0	EA.	120	360	480
1090	Oak				
1100	3-0 x 7-0	EA.	120	310	430
1120	6-0 x 7-0	"	120	360	480
1200	Pine				
1240	2-6 x 7-0	EA.	97.00	180	277
1300	3-0 x 7-0	"	97.00	190	287
1320	3-4 x 7-0	"	97.00	210	307
1340	6-0 x 7-0	"	160	230	390
3000	Fire-rated wood jambs				
3010	20-Minute, positive or neutral pressure, 3/4"	L.F.	4.02	8.80	12.82
3020	45-60-minute	"	4.02	9.90	13.92
3030	90-minute	"	4.02	11.00	15.02
3040	120-minute	"	4.02	13.25	17.27
08300.10	**SPECIAL DOORS**				
1000	Vault door and frame, class 5, steel	EA.	480	6,160	6,640
1480	Overhead door, coiling insulated				
1500	Chain gear, no frame, 12' x 12'	EA.	600	3,410	4,010
2000	Aluminum, bronze glass panels, 12-9 x 13-0	"	480	3,850	4,330
2200	Garage door, flush insulated metal, primed, 9-0 x 7-0	"	160	1,040	1,200
3000	Sliding metal fire doors, motorized, fusible link, 3 hr.				
3040	3-0 x 6-8	EA.	970	3,900	4,870
3060	3-8 x 6-8	"	970	3,900	4,870
3080	4-0 x 8-0	"	970	4,240	5,210
3100	5-0 x 8-0	"	970	4,280	5,250
3200	Metal clad doors, including electric motor				
3210	Light duty				
3220	Minimum	S.F.	8.04	53.00	61.04
3240	Maximum	"	19.25	83.00	102
3250	Heavy duty				
3260	Minimum	S.F.	24.25	83.00	107
3280	Maximum	"	30.25	140	170
3500	Counter doors, (roll-up shutters), standard, manual				
3510	Opening, 4' high				
3520	4' wide	EA.	400	1,300	1,700
3540	6' wide	"	400	1,760	2,160
3560	8' wide	"	440	1,980	2,420
3580	10' wide	"	600	2,200	2,800
3590	14' wide	"	600	2,750	3,350
3595	6' high				
3600	4' wide	EA.	400	1,540	1,940
3620	6' wide	"	440	2,010	2,450
3630	8' wide	"	480	2,200	2,680
3640	10' wide	"	600	2,480	3,080
3650	14' wide	"	690	2,810	3,500
3660	For stainless steel, add to material, 40%				
3670	For motor operator, add	EA.			1,520
3800	Service doors, (roll up shutters), standard, manual				
3810	Opening				
3820	8' high x 8' wide	EA.	270	1,650	1,920
3840	12' high x 12' wide	"	600	2,310	2,910

		UNIT	LABOR	MAT.	TOTAL
08300.10	**SPECIAL DOORS, Cont'd...**				
3860	16' high x 14' wide	EA.	800	4,240	5,040
3870	20' high x 14' wide	"	1,210	5,720	6,930
3880	24' high x 16' wide	"	1,070	7,810	8,880
3890	For motor operator				
3900	Up to 12-0 x 12-0, add	EA.			1,550
3920	Over 12-0 x 12-0, add	"			1,980
4040	Roll-up doors				
4050	13-0 high x 14-0 wide	EA.	690	1,400	2,090
4060	12-0 high x 14-0 wide	"	690	1,790	2,480
5200	Sectional wood overhead doors, frames not included				
5220	Commercial grade, heavy duty, 1-3/4" thick, manual				
5240	8' x 8'	EA.	400	1,030	1,430
5260	10' x 10'	"	440	1,480	1,920
5280	12' x 12'	"	480	1,790	2,270
5290	Chain hoist				
5300	12' x 16' high	EA.	800	2,860	3,660
5320	14' x 14' high	"	600	3,140	3,740
5340	20' x 8' high	"	970	2,690	3,660
5360	16' high	"	1,210	6,050	7,260
5990	Sectional metal overhead doors, complete				
6000	Residential grade, manual				
6020	9' x 7'	EA.	190	710	900
6040	16' x 7'	"	240	1,270	1,510
6100	Commercial grade				
6120	8' x 8'	EA.	400	820	1,220
6140	10' x 10'	"	440	1,090	1,530
6160	12' x 12'	"	480	1,810	2,290
6180	20' x 14', with chain hoist	"	970	4,350	5,320
6400	Sliding glass doors				
6420	Tempered plate glass, 1/4" thick				
6430	6' wide				
6440	Economy grade	EA.	160	1,010	1,170
6450	Premium grade	"	160	1,160	1,320
6455	12' wide				
6460	Economy grade	EA.	240	1,430	1,670
6465	Premium grade	"	240	2,150	2,390
6470	Insulating glass, 5/8" thick				
6475	6' wide				
6480	Economy grade	EA.	160	1,250	1,410
6500	Premium grade	"	160	1,610	1,770
6505	12' wide				
6510	Economy grade	EA.	240	1,560	1,800
6515	Premium grade	"	240	2,500	2,740
6520	1" thick				
6525	6' wide				
6530	Economy grade	EA.	160	1,580	1,740
6540	Premium grade	"	160	1,820	1,980
6545	12' wide				
6550	Economy grade	EA.	240	2,440	2,680
6560	Premium grade	"	240	3,580	3,820
6600	Added costs				
6620	Custom quality, add to material, 30%				

DIVISION # 08 DOORS AND WINDOWS

		UNIT	LABOR	MAT.	TOTAL
08300.10	**SPECIAL DOORS, Cont'd...**				
6630	Tempered glass, 6' wide, add	S.F.			4.01
6880	Residential storm door				
6900	Minimum	EA.	80.00	160	240
6920	Average	"	80.00	220	300
6940	Maximum	"	120	480	600
08410.10	**STOREFRONTS**				
0135	Storefront, aluminum and glass				
0140	Minimum	S.F.	6.71	20.00	26.71
0150	Average	"	7.67	29.75	37.42
0160	Maximum	"	8.95	53.00	61.95
1020	Entrance doors, premium, incl. glass, closers, panic dev.,etc.				
1030	1/2" thick glass				
1040	3' x 7'	EA.	450	3,190	3,640
1060	6' x 7'	"	670	5,450	6,120
1065	3/4" thick glass				
1070	3' x 7'	EA.	450	3,300	3,750
1080	6' x 7'	"	670	5,500	6,170
1085	1" thick glass				
1090	3' x 7'	EA.	450	3,580	4,030
1100	6' x 7'	"	670	6,330	7,000
1150	Revolving doors				
1151	7' diameter, 7' high				
1160	Minimum	EA.	5,280	24,920	30,200
1170	Average	"	8,450	31,350	39,800
1180	Maximum	"	10,570	40,460	51,030
08510.10	**STEEL WINDOWS**				
0100	Steel windows, primed				
1000	Casements				
1010	Operable				
1020	Minimum	S.F.	3.16	37.00	40.16
1040	Maximum	"	3.58	55.00	58.58
1060	Fixed sash	"	2.68	29.50	32.18
1080	Double hung	"	2.98	55.00	57.98
1100	Industrial windows				
1120	Horizontally pivoted sash	S.F.	3.58	52.00	55.58
1130	Fixed sash	"	2.98	40.50	43.48
1135	Security sash				
1140	Operable	S.F.	3.58	65.00	68.58
1150	Fixed	"	2.98	57.00	59.98
1155	Picture window	"	2.98	27.75	30.73
1160	Projecting sash				
1170	Minimum	S.F.	3.35	48.00	51.35
1180	Maximum	"	3.35	59.00	62.35
1930	Mullions	L.F.	2.68	12.50	15.18
08520.10	**ALUMINUM WINDOWS**				
0110	Jalousie				
0120	3-0 x 4-0	EA.	67.00	310	377
0140	3-0 x 5-0	"	67.00	370	437
0220	Fixed window				
0240	6 sf to 8 sf	S.F.	7.67	15.75	23.42
0250	12 sf to 16 sf	"	5.96	14.00	19.96

DIVISION # 08 DOORS AND WINDOWS

		UNIT	LABOR	MAT.	TOTAL
08520.10	**ALUMINUM WINDOWS, Cont'd...**				
0255	Projecting window				
0260	6 sf to 8 sf	S.F.	13.50	35.00	48.50
0270	12 sf to 16 sf	"	8.95	31.50	40.45
0275	Horizontal sliding				
0280	6 sf to 8 sf	S.F.	6.71	22.75	29.46
0290	12 sf to 16 sf	"	5.37	21.00	26.37
1140	Double hung				
1160	6 sf to 8 sf	S.F.	10.75	31.50	42.25
1180	10 sf to 12 sf	"	8.95	28.00	36.95
3010	Storm window, 0.5 cfm, up to				
3020	60 u.i. (united inches)	EA.	26.75	74.00	101
3060	80 u.i.	"	26.75	84.00	111
3080	90 u.i.	"	29.75	86.00	116
3100	100 u.i.	"	29.75	88.00	118
3110	2.0 cfm, up to				
3120	60 u.i.	EA.	26.75	95.00	122
3160	80 u.i.	"	26.75	98.00	125
3180	90 u.i.	"	29.75	110	140
3200	100 u.i.	"	29.75	110	140
08600.10	**WOOD WINDOWS**				
0980	Double hung				
0990	24" x 36"				
1000	Minimum	EA.	48.25	200	248
1002	Average	"	60.00	290	350
1004	Maximum	"	80.00	390	470
1010	24" x 48"				
1020	Minimum	EA.	48.25	230	278
1022	Average	"	60.00	330	390
1024	Maximum	"	80.00	470	550
1030	30" x 48"				
1040	Minimum	EA.	54.00	240	294
1042	Average	"	69.00	340	409
1044	Maximum	"	97.00	490	587
1050	30" x 60"				
1060	Minimum	EA.	54.00	260	314
1062	Average	"	69.00	420	489
1064	Maximum	"	97.00	530	627
1160	Casement				
1180	1 leaf, 22" x 38" high				
1220	Minimum	EA.	48.25	290	338
1222	Average	"	60.00	350	410
1224	Maximum	"	80.00	410	490
1230	2 leaf, 50" x 50" high				
1240	Minimum	EA.	60.00	780	840
1242	Average	"	80.00	1,010	1,090
1244	Maximum	"	120	1,160	1,280
1250	3 leaf, 71" x 62" high				
1260	Minimum	EA.	60.00	1,280	1,340
1262	Average	"	80.00	1,310	1,390
1264	Maximum	"	120	1,560	1,680
1290	5 leaf, 119" x 75" high				

		UNIT	LABOR	MAT.	TOTAL
08600.10	**WOOD WINDOWS, Cont'd...**				
1300	Minimum	EA.	69.00	2,200	2,269
1302	Average	"	97.00	2,370	2,467
1304	Maximum	"	160	3,040	3,200
1360	Picture window, fixed glass, 54" x 54" high				
1400	Minimum	EA.	60.00	450	510
1422	Average	"	69.00	510	579
1424	Maximum	"	80.00	910	990
1430	68" x 55" high				
1440	Minimum	EA.	60.00	820	880
1442	Average	"	69.00	940	1,009
1444	Maximum	"	80.00	1,230	1,310
1480	Sliding, 40" x 31" high				
1520	Minimum	EA.	48.25	270	318
1522	Average	"	60.00	410	470
1524	Maximum	"	80.00	490	570
1530	52" x 39" high				
1540	Minimum	EA.	60.00	340	400
1542	Average	"	69.00	500	569
1544	Maximum	"	80.00	560	640
1550	64" x 72" high				
1560	Minimum	EA.	60.00	520	580
1562	Average	"	80.00	830	910
1564	Maximum	"	97.00	920	1,017
1760	Awning windows				
1780	34" x 21" high				
1800	Minimum	EA.	48.25	280	328
1822	Average	"	60.00	330	390
1824	Maximum	"	80.00	370	450
1880	48" x 27" high				
1900	Minimum	EA.	54.00	360	414
1902	Average	"	69.00	420	489
1904	Maximum	"	97.00	490	587
1920	60" x 36" high				
1940	Minimum	EA.	60.00	370	430
1942	Average	"	80.00	660	740
1944	Maximum	"	97.00	740	837
8000	Window frame, milled				
8010	Minimum	L.F.	9.65	5.28	14.93
8020	Average	"	12.00	5.85	17.85
8030	Maximum	"	16.00	8.83	24.83
08710.10	**HINGES**				
1200	Hinges				
1250	3 x 3 butts, steel, interior, plain bearing	PAIR			20.75
1260	4 x 4 butts, steel, standard	"			30.50
1270	5 x 4-1/2 butts, bronze/s. steel, heavy duty	"			79.00
1290	Pivot hinges				
1300	Top pivot	EA.			76.00
1310	Intermediate pivot	"			81.00
1320	Bottom pivot	"			160

DIVISION # 08 DOORS AND WINDOWS

		UNIT	LABOR	MAT.	TOTAL
08710.20	**LOCKSETS**				
1280	Latchset, heavy duty				
1300	Cylindrical	EA.	30.25	160	190
1320	Mortise	"	48.25	160	208
1325	Lockset, heavy duty				
1330	Cylindrical	EA.	30.25	330	360
1350	Mortise	"	48.25	390	438
2200	Preassembled locks and latches, brass				
2220	Latchset, passage or closet latch	EA.	40.25	250	290
2225	Lockset				
2230	Privacy (bath or bathroom)	EA.	40.25	300	340
2240	Entry lock	"	40.25	430	470
2285	Lockset				
2290	Privacy (bath or bedroom)	EA.	40.25	190	230
2300	Entry lock	"	40.25	220	260
08710.30	**CLOSERS**				
2600	Door closers				
2610	Standard	EA.	60.00	220	280
2620	Heavy duty	"	60.00	260	320
08710.40	**DOOR TRIM**				
1600	Panic device				
1610	Mortise	EA.	120	780	900
1620	Vertical rod	"	120	1,170	1,290
1630	Labeled, rim type	"	120	810	930
1640	Mortise	"	120	1,040	1,160
1650	Vertical rod	"	120	1,130	1,250
2300	Door plates				
2305	Kick plate, aluminum, 3 beveled edges				
2310	10" x 28"	EA.	24.25	22.00	46.25
2340	10" x 38"	"	24.25	28.50	52.75
2350	Push plate, 4" x 16"				
2360	Aluminum	EA.	9.65	22.00	31.65
2371	Bronze	"	9.65	84.00	93.65
2380	Stainless steel	"	9.65	67.00	76.65
2385	Armor plate, 40" x 34"	"	19.25	77.00	96.25
2388	Pull handle, 4" x 16"				
2390	Aluminum	EA.	9.65	95.00	105
2400	Bronze	"	9.65	180	190
2420	Stainless steel	"	9.65	140	150
2425	Hasp assembly				
2430	3"	EA.	8.04	3.52	11.56
2440	4-1/2"	"	10.75	4.40	15.15
2450	6"	"	13.75	6.98	20.73
08710.60	**WEATHERSTRIPPING**				
0100	Weatherstrip, head and jamb, metal strip, neoprene bulb				
0140	Standard duty	L.F.	2.68	4.93	7.61
0160	Heavy duty	"	3.01	5.48	8.49
3980	Spring type				
4000	Metal doors	EA.	120	55.00	175
4010	Wood doors	"	160	55.00	215
4020	Sponge type with adhesive backing	"	48.25	51.00	99.25
4500	Thresholds				

		UNIT	LABOR	MAT.	TOTAL
08710.60	**WEATHERSTRIPPING, Cont'd...**				
4510	Bronze	L.F.	12.00	53.00	65.00
4515	Aluminum				
4520	Plain	L.F.	12.00	24.75	36.75
4525	Vinyl insert	"	12.00	29.75	41.75
4530	Aluminum with grit	"	12.00	28.50	40.50
4533	Steel				
4535	Plain	L.F.	12.00	29.75	41.75
4540	Interlocking	"	40.25	39.50	79.75
08810.10	**GLAZING**				
0800	Sheet glass, 1/8" thick	S.F.	2.98	7.70	10.68
1020	Plate glass, bronze or grey, 1/4" thick	"	4.88	11.25	16.13
1040	Clear	"	4.88	8.80	13.68
1060	Polished	"	4.88	10.50	15.38
1980	Plexiglass				
2000	1/8" thick	S.F.	4.88	3.65	8.53
2020	1/4" thick	"	2.98	6.50	9.48
3000	Float glass, clear				
3010	3/16" thick	S.F.	4.47	5.25	9.72
3020	1/4" thick	"	4.88	6.12	11.00
3040	3/8" thick	"	6.71	12.25	18.96
3100	Tinted glass, polished plate, twin ground				
3120	3/16" thick	S.F.	4.47	8.22	12.69
3130	1/4" thick	"	4.88	8.22	13.10
3140	3/8" thick	"	6.71	13.25	19.96
5000	Insulated glass, bronze or gray				
5020	1/2" thick	S.F.	8.95	15.25	24.20
5040	1" thick	"	13.50	18.25	31.75
5100	Spandrel glass, polished bronze/grey, 1 side, 1/4" thick	"	4.88	12.25	17.13
6000	Tempered glass (safety)				
6010	Clear sheet glass				
6020	1/8" thick	S.F.	2.98	8.41	11.39
6030	3/16" thick	"	4.13	10.25	14.38
6040	Clear float glass				
6050	1/4" thick	S.F.	4.47	8.76	13.23
6060	5/16" thick	"	5.37	15.75	21.12
6070	3/8" thick	"	6.71	19.25	25.96
6080	1/2" thick	"	8.95	26.25	35.20
6800	Insulating glass, two lites, clear float glass				
6840	1/2" thick	S.F.	8.95	12.00	20.95
6850	5/8" thick	"	10.75	13.75	24.50
6860	3/4" thick	"	13.50	15.25	28.75
6870	7/8" thick	"	15.25	16.00	31.25
6880	1" thick	"	18.00	17.00	35.00
6885	Glass seal edge				
6890	3/8" thick	S.F.	8.95	10.00	18.95
6895	Tinted glass				
6900	1/2" thick	S.F.	8.95	20.50	29.45
6910	1" thick	"	18.00	22.00	40.00
6920	Tempered, clear				
6930	1" thick	S.F.	18.00	40.25	58.25
7200	Plate mirror glass				

DIVISION # 08 DOORS AND WINDOWS

		UNIT	LABOR	MAT.	TOTAL
08810.10	**GLAZING, Cont'd...**				
7205	1/4" thick				
7210	15 sf	S.F.	5.37	10.50	15.87
7220	Over 15 sf	"	4.88	9.65	14.53
9650	Sand-Blasted Glass				
9660	3/16" Float glass, full sandblast, no custom decoration	S.F.	4.47	10.50	14.97
9670	3/8" Float glass, full sandblast, no custom decoration	"	5.37	12.25	17.62
9680	Beveled glass				
9690	commercial standard grade	S.F.	5.37	130	135
08910.10	**GLAZED CURTAIN WALLS**				
1000	Curtain wall, aluminum system, framing sections				
1005	2" x 3"				
1010	Jamb	L.F.	4.47	11.00	15.47
1020	Horizontal	"	4.47	11.25	15.72
1030	Mullion	"	4.47	15.00	19.47
1035	2" x 4"				
1040	Jamb	L.F.	6.71	15.00	21.71
1060	Horizontal	"	6.71	15.50	22.21
1070	Mullion	"	6.71	15.00	21.71
1080	3" x 5-1/2"				
1090	Jamb	L.F.	6.71	19.75	26.46
1100	Horizontal	"	6.71	22.00	28.71
1110	Mullion	"	6.71	20.00	26.71
1115	4" corner mullion	"	8.95	26.25	35.20
1120	Coping sections				
1130	1/8" x 8"	L.F.	8.95	27.50	36.45
1140	1/8" x 9"	"	8.95	27.75	36.70
1150	1/8" x 12-1/2"	"	10.75	28.50	39.25
1160	Sill section				
1170	1/8" x 6"	L.F.	5.37	27.25	32.62
1180	1/8" x 7"	"	5.37	27.50	32.87
1190	1/8" x 8-1/2"	"	5.37	28.00	33.37
1200	Column covers, aluminum				
1210	1/8" x 26"	L.F.	13.50	27.25	40.75
1220	1/8" x 34"	"	14.25	27.50	41.75
1230	1/8" x 38"	"	14.25	27.75	42.00
1500	Doors				
1600	Aluminum framed, standard hardware				
1620	Narrow stile				
1630	2-6 x 7-0	EA.	270	610	880
1640	3-0 x 7-0	"	270	620	890
1660	3-6 x 7-0	"	270	640	910
1700	Wide stile				
1720	2-6 x 7-0	EA.	270	1,060	1,330
1730	3-0 x 7-0	"	270	1,130	1,400
1750	3-6 x 7-0	"	270	1,220	1,490
1800	Flush panel doors, to match adjacent wall panels				
1810	2-6 x 7-0	EA.	340	890	1,230
1820	3-0 x 7-0	"	340	940	1,280
1840	3-6 x 7-0	"	340	970	1,310
2100	Wall panel, insulated				
2120	"U"=.08	S.F.	4.47	11.25	15.72

		UNIT	LABOR	MAT.	TOTAL
08910.10	**GLAZED CURTAIN WALLS, Cont'd...**				
2140	"U"=.10	S.F.	4.47	10.50	14.97
2160	"U"=.15	"	4.47	9.54	14.01
3000	Window wall system, complete				
3010	Minimum	S.F.	5.37	19.75	25.12
3030	Average	"	5.96	31.50	37.46
3050	Maximum	"	7.67	73.00	80.67
4860	Added costs				
4870	For bronze, add 20% to material				
4880	For stainless steel, add 50% to material				

		UNIT	COST
08999.10	**DOORS**		
0900	HOLLOW METAL		
1000	CUSTOM FRAMES (16 ga)		
1010	2'6" x 6'8" x 4 3/4"	EA.	332
1020	2'8" x 6'8" x 4 3/4"	"	346
1030	3'0" x 6'8" x 4 3/4"	"	370
1040	3'4" x 6'8" x 4 3/4"	"	390
2000	Add for A, B or C Label	"	41.75
2010	Add for Side Lights	"	193
2020	Add for Frames over 7'0" - Height	"	77.25
2030	Add for Frames over 6 3/4" - Width	"	67.25
3000	CUSTOM DOORS (1 3/8" or 1 3/4") (18 ga)		
3010	2'6" x 6'8" x 1 3/4"	EA.	390
3020	2'8" x 6'8" x 1 3/4"	"	400
3030	3'0" x 6'8" x 1 3/4"	"	420
3040	3'4" x 6'8" x 1 3/4"	"	430
4000	Add for 16 ga	"	77.25
4010	Add for B and C Label	"	42.00
4020	Add for Vision Panels or Lights	"	95.50
5000	STOCK FRAMES (16 ga)		
5010	2'6" x 6'8" or 7'0" x 4 3/4"	EA.	220
5020	2'8" x 6'8" or 7'0" x 4 3/4"	"	230
5030	3'0" x 6'8" or 7'0" x 4 3/4"	"	240
5040	3'4" x 6'8" or 7'0" x 4 3/4"	"	260
5050	Add for A, B or C Label	"	40.25
6000	STOCK DOORS (1 3/8" or 1 3/4" - 18 ga)		
6010	2'6" x 6'8" or 7'0"	EA.	360
6020	2'8" x 6'8" or 7'0"	"	370
6030	3'0" x 6'8" or 7'0"	"	390
6040	3'4" x 6'8" or 7'0"	"	410
6050	Add for B or C Label	"	33.35
08999.20	**WOOD DOORS (incl. butts and locksets)**		
1000	FLUSH DOORS - No Label - Paint Grade		
2000	Birch - 7 Ply 2'4" x 6'8" x 1 3/8" -Hollow Core	EA.	220
2010	2'6" x 6'8" x 1 3/8"	"	220
2020	2'8" x 6'8" x 1 3/8"	"	240
2030	3'0" x 6'8" x 1 3/8"	"	230
2040	2'4" x 6'8" x 1 3/4"	"	240
2050	2'6" x 6'8" x 1 3/4"	"	240
2060	2'8" x 6'8" x 1 3/4"	"	250
2070	3'0" x 6'8" x 1 3/4"	"	260
2080	3'4" x 6'8" x 1 3/4"	"	260
2090	3'6" x 6'8" x 1 3/4"	"	260
3000	Birch - 7 Ply 2'4" x 6'8" x 1 3/8" -Solid Core	"	260
3010	2'6" x 6'8" x 1 3/8"	"	260
3020	2'8" x 6'8" x 1 3/8"	"	300
3030	3'0" x 6'8" x 1 3/8"	"	300
3040	2'4" x 6'8" x 1 3/4"	"	300
3050	2'6" x 6'8" x 1 3/4"	"	300
3060	2'8" x 6'8" x 1 3/4"	"	325
3070	3'0" x 6'8" x 1 3/4"	"	325
3080	3'4" x 6'8" x 1 3/4"	"	335

		UNIT	COST
08999.20	**WOOD DOORS (Incl. butts and locksets), Cont'd...**		
3090	3'6" x 6'8" x 1 3/4"E	EA.	390
4000	Add for Stain Grade	"	42.50
4010	Add for 5 Ply	"	91.50
4020	Add for Jamb & Trim - Solid - Knock Down	"	94.50
4030	Add for Jamb & Trim - Veneer - Knock Down	"	62.50
4040	Add for Red Oak - Rotary Cut	"	38.00
4050	Add for Red Oak - Plain Sliced	"	51.25
5000	Deduct for Lauan	"	30.12
5010	Add for Architectural Grade - 7 Ply	"	88.75
5020	Add for Vinyl Overlay	"	68.75
5030	Add for 7' - 0" Doors	"	30.12
5040	Add for Lite Cutouts with Metal Frame	"	100
5050	Add for Wood Louvres	"	152
5060	Add for Transom Panels & Side Panels	"	193
6000	LABELED - 1 3/4" - Paint Grade		
6010	2'6" x 6'8" - Birch - 20 Min.	EA.	335
6020	2'8" x 6'8"	"	315
6030	3'0" x 6'8"	"	325
6040	3'4" x 6'8"	"	370
6050	3'6" x 6'8"	"	420
7000	2'6" x 6'8" - Birch - 45 Min	"	350
7010	2'8" x 6'8"	"	370
7020	3'0" x 6'8"	"	380
7030	3'4" x 6'8"	"	400
7040	3'6" x 6'8"	"	410
8000	2'6" x 6'8" - Birch - 60 Min	"	350
8010	2'8" x 6'8"	"	370
8020	3'0" x 6'8"	"	380
8030	3'4" x 6'8"	"	400
8040	3'6" x 6'8"	"	315
9000	PANEL DOORS		
9010	Exterior - 1 3/4" -Pine		
9020	2'8" x 6'8"	EA.	530
9030	3'0" x 6'8"	"	540
9040	Interior - 1 3/8" -Pine		
9050	2'6" x 6'8"	EA.	390
9060	2'8" x 6'8"	"	420
9070	3'0" x 6'8"	"	450
9080	Add for Sidelights	"	360
9140	Exterior - 1 3/4"- Birch/ Oak		
9150	2'8" x 6'8"	EA.	740
9160	3'0" x 6'8"	"	780
9170	Interior - 1 3/8"		
9180	2'6" x 6'8"	EA.	540
9190	2'8" x 6'8"	"	560
9200	3'0" x 6'8"	"	600
9210	Add for Sidelights	"	550
9212	LOUVERED DOORS - 1 3/8" -Pine		
9213	2'0" x 6'8"	EA.	300
9214	2'6" x 6'8"	"	340
9215	2'8" x 6'8"	"	360
9216	3'0" x 6'8"	"	380

		UNIT	COST
08999.20	**WOOD DOORS (Incl. butts and locksets), Cont'd...**		
9220	Birch/Oak - 1 3/8"		
9230	2'0" x 6'8"	EA.	710
9240	2'6" x 6'8"	"	720
9250	2'8" x 6'8"	"	770
9260	3'0" x 6'8"	"	780
9270	BI-FOLD with hdwr. 1 3/8" - Flush		
9280	Birch/Oak, 2'0" x 6'8"	EA.	220
9290	2'4" x 6'8"	"	245
9300	2'6" x 6'8"	"	265
9320	Lauan, 2'0" x 6'8"	"	220
9330	2'4" x 6'8"	"	230
9340	2'6" x 6'8"	"	240
9350	Add for Prefinished	"	28.12
9360	CAFE DOORS - 1 1/8" - Pair		
9370	2'6" x 3'8"	EA.	450
9380	2'8" x 3'8"	"	460
9390	3'0" x 3'8"	"	480
9400	FRENCH DOORS - 1 3/8" -Pine		
9410	2'6" x 3'8"	EA.	500
9420	2'8" x 3'8"	"	510
9430	3'0" x 3'8"	"	530
9440	2'6" x 3'8" -Birch/Oak	"	780
9450	2'8" x 3'8"	"	790
9460	3'0" x 3'8"	"	800
9470	DUTCH DOORS		
9480	2'6" x 3'8"	EA.	880
9490	2'8" x 3'8"	"	970
9500	3'0" x 3'8"	"	1,230
9510	PREHUNG DOORS (INCL. TRIM)		
9520	Exterior, entrance - 1 3/4"		
9530	Panel - 2'8" x 6'8"	EA.	580
9540	3'0" x 6'8"	"	640
9550	3'0" x 7'0"	"	700
9560	Add for Insulation	"	260
9570	Add for Side Lights	"	250
9580	Interior - 1 3/8"		
9590	Flush, H.C. - 2'6" x 6'8" -Birch/Oak	EA.	420
9600	2'8" x 6'8"	"	430
9610	3'0" x 6'8"Each	"	450
9620	Flush, H.C. - 2'6" x 6'8" -Lauan	"	325
9630	2'8" x 6'8"	"	360
9640	3'0" x 6'8"Each	"	360
9650	Add for Int. Panel Door	"	295
08999.30	**SPECIAL DOORS**		
1000	BI-FOLDING - PREHUNG, 1 3/8"		
1020	Wood - 2 Door - 2'0" x 6'8" -Oak/flush	EA.	325
1030	2'6" x 6'8"	"	370
1040	3'0" x 6'8"	"	390
1050	4 Door - 4'0" x 6'8"	"	420
1060	5'0" x 6'8"	"	430
1070	6'0" x 6'8"	"	500

		UNIT	COST
08999.30	**SPECIAL DOORS, Cont'd...**		
1080	Add for Prefinished	EA.	39.00
1090	Add for Jambs and Casings	"	95.50
2000	Wood - 2 Door - 2'0" x 6'8" Pine/Panel	"	640
2010	2'6" x 6'8"	"	650
2020	3'0" x 6'8"	"	760
2030	4 Door - 4'0" x 6'8"	"	960
2040	5'0" x 6'8"	"	1,040
2050	6'0" x 6'8"	"	1,250
3000	Metal - 2 Door - 2'0" x 6'8"	"	203
3010	2'6" x 6'8"	"	203
3020	3'0" x 6'8"	"	250
3030	4 Door - 4'0" x 6'8"	"	270
3040	5'0" x 6'8"	"	295
3050	6'0" x 6'8"	"	315
3060	Add for Plastic Overlay	"	46.25
3070	Add for Louvre or Decorative Type	"	240
4000	Leaded Mirror (Based on 2-Panel Unit)		
4010	4'0" x 6'8"	EA.	930
4020	5'0" x 6'8"	"	1,070
4030	6'0" x 6'8"	"	1,070
7000	ROLLING - DOORS & GRILLES		
7010	Doors - 8' x 8'	EA.	3,710
7020	10' x 10'	"	3,890
7030	Grilles - 6'-8" x 3'-2"	"	1,290
7040	6'-8" x 4'-2"	"	1,450
8000	SHOWER DOORS - 28" x 66"	"	370
9000	SLIDING OR PATIO DOORS		
9010	Metal (Aluminum) - Including Glass Thresholds & Screen		
9020	Opening 8'0" x 6'8"	EA.	2,060
9030	8'0" x 8'0"	"	2,320
9040	Wood		
9050	Vinyl Clad 6'0" x 6'10"	EA.	2,570
9060	8'0" x 6'10"	"	2,820
9070	Pine - Prefinished 6'0" x 6'10"	"	2,540
9080	8'0" x 6'10"	"	2,830
9090	Add for Grilles	"	420
9100	Add for Triple Glazing	"	270
9110	Add for Screen	"	193
9120	SOUND REDUCTION - Metal	"	2,460
9130	Wood	"	1,620
9200	VAULT DOORS - 6'6" x 2'2" with Frame - 2 hour	"	4,230
08999.50	**METAL WINDOWS**		
1000	EACH, 2'4" x 4'6" - ALUMINUM WINDOWS		
1010	Casement and Awning	EA.	460
1020	Sliding or Horizontal	"	373
1030	Double and Single-Hung	"	403
1040	or Vertical Sliding		
1050	Projected	EA.	460
1060	Add for Screens	"	66.18
1070	Add for Storms	"	116
1080	Add for Insulated Glass	"	101

		UNIT	COST
08999.50	**METAL WINDOWS, Cont'd...**		
2000	ALUMINUM SASH		
2010	Casement	EA.	423
2020	Sliding	"	373
2030	Single-Hung	"	373
2040	Projected	"	312
2050	Fixed	"	322
3000	STEEL WINDOWS		
3010	Double-Hung	EA.	655
3020	Projected	"	625
4000	STEEL SASH		
4010	Casement	EA.	520
4020	Double-Hung	"	570
4030	Projected	"	440
4040	Fixed	"	440
5000	By S.F., 2'4" x 4'6" - ALUMINUM WINDOWS		
5010	Casement and Awning	S.F.	45.78
5020	Sliding or Horizontal	"	35.94
5030	Double and Single-Hung	"	40.53
5040	or Vertical Sliding	"	
5050	Projected	"	45.62
5060	Add for Screens	"	6.48
5070	Add for Storms	"	11.78
5080	Add for Insulated Glass	"	9.85
6000	ALUMINUM SASH		
6010	Casement	S.F.	42.81
6020	Sliding	"	36.19
6030	Single-Hung	"	36.19
6040	Projected	"	30.73
6050	Fixed	"	31.89
7000	STEEL WINDOWS		
7010	Double-Hung	S.F.	67.52
7020	Projected	"	64.12
8000	STEEL SASH		
8010	Casement	S.F.	52.10
8020	Double-Hung	"	64.12
8030	Projected	"	44.63
8040	Fixed	"	46.08
08999.60	**WOOD WINDOWS**		
1000	BASEMENT OR UTILITY		
1020	Prefinished with Screen, 2'8" x 1'4"	EA.	252
1030	2'8" x 2'0"	"	293
1040	Add for Double Glazing or Storm	"	64.75
2000	CASEMENT OR AWNING		
2010	Operating Units - Insulating Glass		
2020	Single 2'4" x 4'0"	EA.	590
2030	2'4" x 5'0"	"	650
3000	Double with Screen 4'0" x 4'0" (2 units)	"	910
3010	4'0" x 5'0" (2 units)	"	1,160
3020	Triple with Screen 6'0" x 4'0" (3 units)	"	1,590
3030	6'0" x 5'0" (3 units)	"	1,600
4000	Fixed Units - Insulating Glass		

		UNIT	COST
08999.60	**WOOD WINDOWS, Cont'd...**		
4010	Single 2'4" x 4'0"	EA.	570
4020	2'4" x 5'0"	"	640
4030	Picture 4'0" x 4'6"	"	780
4040	4'0" x 6'0"	"	930
5000	DOUBLE HUNG - Insul. Glass - 2'6" x 3'6"	"	610
5010	2'6" x 4'2"	"	610
5020	3'2" x 3'6"	"	630
5030	3'2" x 4'2"	"	700
5040	Add for Triple Glazing	"	97.50
6000	GLIDER - & Insul. Glass with Screen - 4'0" x 3'6"	"	1,410
6010	5'0" x 4'0"	"	1,590
8000	PICTURE WINDOWS		
8010	9'6" x 4'10"	EA.	2,960
8020	9'6" x 5'6"	"	2,250
9000	CASEMENT - Insulating Glass		
9010	30° - 5'10" x 4'2" - (3 units)	EA.	2,140
9015	7'10" x 4'2" - (4 units)	"	2,600
9020	5'10" x 5'2" - (3 units)	"	2,240
9030	7'10" x 5'2" - (4 units)	"	2,550
9040	45° - 5'4" x 4'2" - (3 units)	"	2,130
9050	7'4" x 4'2" - (4 units)	"	2,350
9060	5'4" x 5'2" - (3 units)	"	2,500
9070	7'4" x 5'2" - (4 units)	"	2,700
9100	CASEMENT BOW WINDOWS - Insulating Glass		
9110	6'2" x 4'2" - (3 units)	EA.	2,140
9120	8'2" x 4'2" - (4 units)	"	2,850
9130	6'2" x 5'2" - (3 units)	"	2,400
9140	8'2" x 5'2" - (4 units)	"	3,150
9150	Add for Triple Glazing - per unit	"	79.50
9160	Add for Bronze Glazing - per unit	"	79.50
9170	Deduct for Primed Only - per unit	"	20.24
9180	Add for Screens - per unit	"	24.84
9200	90° BOX BAY WINDOWS - Insulating Glass 4'8" x 4'2"	"	2,850
9210	6'8" x 4'2"	"	3,630
9220	6'8" x 5'2"	"	3,770
9300	ROOF WINDOWS 1'10" x 3'10"-Fixed	"	840
9310	2'4" x 3'10"	"	950
9320	3'8" x 3'10"	"	1,170
9400	ROOF WINDOWS 1'10" x 3'10"-Movable	"	1,110
9410	2'4" x 3'10"	"	1,370
9420	3'8" x 3'10"	"	1,580
9500	CIRCULAR TOPS & ROUNDS - 4'0"	"	900
9510	6'0"	"	1,980
08999.70	**SPECIAL WINDOWS**		
1000	LIGHT-PROOF WINDOWS	S.F.	52.85
2000	PASS WINDOWS	"	40.79
3000	DETENTION WINDOWS	"	55.25
4000	VENETIAN BLIND WINDOWS (ALUMINUM)	"	46.20
5000	SOUND-CONTROL WINDOWS	"	45.03

		UNIT	COST
08999.80	**DOOR AND WINDOW ACCESSORIES**		
1000	STORMS AND SCREENS		
1010	Windows		
1020	Screen Only - Wood - 3' x 5'	EA.	147
1030	Aluminum - 3' x 5'	"	160
1040	Storm & Screen Combination - Aluminum	"	183
1050	Doors	"	
1060	Screen Only - Wood	"	340
1070	Aluminum - 3' x 6' - 8'	"	340
1080	Storm & Screen Combination - Aluminum	"	400
1090	Wood 1 1/8"	"	455
2000	DETENTION SCREENS		
2010	Example: 4' 0" x 7' 0"	EA.	960
3000	DOOR OPENING ASSEMBLIES		
3010	Floor or Overhead Electric Eye Units		
3020	Swing - Single 3' x 7' Door - Hydraulic	EA.	5,600
3030	Double 6' x 7' Door - Hydraulic	"	8,840
3040	Sliding - Single 3' x 7' Door - Hydraulic	"	6,870
3050	Double 5' x 7' Doors- Hydraulic	"	8,810
3060	Industrial Doors - 10' x 8'	"	9,930
4000	SHUTTERS		
4010	16" x 1 1/8" x 48"	EA.	158
4020	16" x 1 1/8" x 60"	"	183
4030	16" x 1 1/8" x 72"	"	213
08999.90	**FINISH HARDWARE**		
1000	BUTT HINGES -Painted		
1010	3" x 3"	EA.	46.75
1020	3 1/2" x 3 1/2"	"	49.25
1030	4" x 4"	"	53.00
1040	4 1/2" x 4 1/2"	"	57.00
1050	4" x 4" Ball Bearing	"	87.00
1060	4 1/2" x 4 1/2" Ball Bearing	"	93.00
2000	Bronze		
2010	3" x 3"	EA.	49.25
2020	3 1/2" x 3 1/2"	"	50.00
2030	4" x 4"	"	53.00
2040	4 1/2" x 4 1/2"	"	61.00
2050	4" x 4" Ball Bearing	"	89.00
2060	4 1/2" x 4 1/2" Ball Bearing	"	100
3000	Chrome		
3010	3" x 3"	EA.	53.00
3020	3 1/2" x 3 1/2"	"	56.00
3030	4" x 4"	"	59.00
3040	4 1/2" x 4 1/2"	"	70.00
3050	4" x 4" Ball Bearing	"	100
3060	4 1/2" x 4 1/2" Ball Bearing	"	110
5000	CLOSERS - SURFACE MOUNTED - 3' - 0" Door	"	255
5010	3' - 4"	"	265
5020	3' - 8"	"	275
5030	4' - 0"	"	350
6000	CLOSERS - CONCEALED - Interior	"	420
6010	ExteriorEach	"	560

		UNIT	COST
08999.90	**FINISH HARDWARE, Cont'd...**		
6020	Add for Fusible Link - Electric	EA.	245
6030	CLOSERS - FLOOR HINGES - Interior	"	630
6040	Exterior	"	950
6050	Add for Hold Open Feature	"	99.50
6060	Add for Double Acting Feature	"	330
7000	DEAD BOLT LOCK - Cylinder - Outside Key	"	215
7010	Cylinder - Double Key	"	310
7020	Flush - Push/ Pull	"	83.50
8000	EXIT DEVICES (PANIC) - Surface	"	920
8010	Mortise Lock	"	1,160
8020	Concealed	"	2,900
8030	Handicap (ADA) Automatic	"	2,130
9000	HINGES, SPRING (PAINTED) - 6" Single Acting	"	157
9010	6" Double Acting	"	190
9101	LATCHSETS -Bronze or Chrome	"	255
9102	Stainless Steel	"	290
9200	LOCKSETS - Mortise - Bronze or Chrome - H.D.	"	350
9210	Mortise - Bronze or Chrome - S.D.	"	275
9220	Stainless Steel	"	450
9230	Cylindrical - Bronze or Chrome	"	310
9240	Stainless Steel	"	360
9300	LEVER HANDICAP - Latch Set	"	360
9310	Lock Set	"	420
9400	PLATES - Kick - 8" x 34" - Aluminum	"	110
9410	Bronze	"	101
9420	Push - 6" x 15" - Aluminum	"	73.25
9430	Bronze	"	104
9440	Push & Pull Combination - Aluminum	"	121
9450	Bronze	"	173
9500	STOPS AND HOLDERS		
9510	Holder - Magnetic (No Electric)	EA.	245
9520	Bumper	"	57.50
9530	Overhead - Bronze, Chrome or Aluminum	"	134
9540	Wall Stops	"	51.00
9550	Floor Stops	"	63.50
08999.91	**WEATHERSTRIPPING**		
1000	ASTRAGALS - Aluminum - 1/8" x 2"	EA.	69.25
1010	Painted Steel	"	62.25
2000	DOORS (WOOD) - Interlocking	"	110
2010	Spring Bronze	"	119
2020	Add for Metal Doors	"	55.00
3000	SWEEPS - 36" WOOD DOORS - Aluminum	"	27.75
3010	Vinyl	"	55.50
4000	THRESHOLDS - Aluminum - 4" x 1/2"	"	66.25
4010	Bronze 4" x 1/2"	"	83.25
4020	5 1/2" x 1/2"	"	88.25
5000	WINDOWS (WOOD) - Interlocking	"	93.50
5010	Spring Bronze	"	111

TABLE OF CONTENTS   PAGE

09110.10 - METAL STUDS   09-2
09205.10 - GYPSUM LATH   09-2
09205.20 - METAL LATH   09-3
09205.60 - PLASTER ACCESSORIES   09-3
09210.10 - PLASTER   09-3
09220.10 - PORTLAND CEMENT PLASTER   09-3
09310.10 - CERAMIC TILE   09-4
09330.10 - QUARRY TILE   09-5
09410.10 - TERRAZZO   09-5
09510.10 - CEILINGS AND WALLS   09-5
09550.10 - WOOD FLOORING   09-6
09630.10 - UNIT MASONRY FLOORING   09-7
09660.10 - RESILIENT TILE FLOORING   09-7
09665.10 - RESILIENT SHEET FLOORING   09-7
09678.10 - RESILIENT BASE AND ACCESSORIES   09-8
09682.10 - CARPET PADDING   09-8
09685.10 - CARPET   09-8
09905.10 - PAINTING PREPARATION   09-8
09910.05 - EXT. PAINTING, SITEWORK   09-9
09910.15 - EXT. PAINTING, BUILDINGS   09-10
09910.35 - INT. PAINTING, BUILDINGS   09-11
09955.10 - WALL COVERING   09-13

		UNIT	LABOR	MAT.	TOTAL
09110.10	**METAL STUDS**				
0060	Studs, non load bearing, galvanized				
0130	3-5/8", 20 ga.				
0140	12" o.c.	S.F.	1.20	0.85	2.05
0142	16" o.c.	"	0.96	0.68	1.64
0144	24" o.c.	"	0.80	0.49	1.29
0170	25 ga.				
0180	12" o.c.	S.F.	1.20	0.57	1.77
0182	16" o.c.	"	0.96	0.46	1.42
0184	24" o.c.	"	0.80	0.35	1.15
0210	6", 20 ga.				
0220	12" o.c.	S.F.	1.50	1.19	2.69
0222	16" o.c.	"	1.20	0.94	2.14
0224	24" o.c.	"	1.00	0.70	1.70
0230	25 ga.				
0240	12" o.c.	S.F.	1.50	0.78	2.28
0242	16" o.c.	"	1.20	0.61	1.81
0244	24" o.c.	"	1.00	0.46	1.46
0980	Load bearing studs, galvanized				
0990	3-5/8", 16 ga.				
1000	12" o.c.	S.F.	1.20	1.55	2.75
1020	16" o.c.	"	0.96	1.43	2.39
1110	18 ga.				
1130	12" o.c.	S.F.	0.80	1.11	1.91
1140	16" o.c.	"	0.96	1.21	2.17
1980	6", 16 ga.				
2000	12" o.c.	S.F.	1.50	2.07	3.57
2001	16" o.c.	"	1.20	1.87	3.07
3000	Furring				
3160	On beams and columns				
3170	7/8" channel	L.F.	3.21	0.51	3.72
3180	1-1/2" channel	"	3.71	0.60	4.31
4460	On ceilings				
4470	3/4" furring channels				
4480	12" o.c.	S.F.	2.01	0.51	2.52
4490	16" o.c.	"	1.93	0.39	2.32
4495	24" o.c.	"	1.72	0.26	1.98
4500	1-1/2" furring channels				
4520	12" o.c.	S.F.	2.19	0.66	2.85
4540	16" o.c.	"	2.01	0.49	2.50
4560	24" o.c.	"	1.85	0.34	2.19
5000	On walls				
5020	3/4" furring channels				
5050	12" o.c.	S.F.	1.60	0.51	2.11
5100	16" o.c.	"	1.50	0.39	1.89
5150	24" o.c.	"	1.42	0.26	1.68
5200	1-1/2" furring channels				
5210	12" o.c.	S.F.	1.72	0.66	2.38
5220	16" o.c.	"	1.60	0.49	2.09
5230	24" o.c.	"	1.50	0.34	1.84

		UNIT	LABOR	MAT.	TOTAL
09205.10	**GYPSUM LATH**				
1070	Gypsum lath, 1/2" thick				
1090	Clipped	S.Y.	2.68	7.20	9.88
1110	Nailed	"	3.01	7.20	10.21
09205.20	**METAL LATH**				
0960	Diamond expanded, galvanized				
0980	2.5 lb., on walls				
1010	Nailed	S.Y.	6.03	3.13	9.16
1030	Wired	"	6.89	3.13	10.02
1040	On ceilings				
1050	Nailed	S.Y.	6.89	3.13	10.02
1070	Wired	"	8.04	3.13	11.17
2230	Stucco lath				
2240	1.8 lb.	S.Y.	6.03	4.96	10.99
2300	3.6 lb.	"	6.03	5.56	11.59
2310	Paper backed				
2320	Minimum	S.Y.	4.82	3.85	8.67
2400	Maximum	"	6.89	6.21	13.10
09205.60	**PLASTER ACCESSORIES**				
0120	Expansion joint, 3/4", 26 ga., galvanized, one piece	L.F.	1.20	0.99	2.19
2000	Plaster corner beads, 3/4", galvanized	"	1.37	0.56	1.93
2020	Casing bead, expanded flange, galvanized	"	1.20	0.56	1.76
2100	Expanded wing, 1-1/4" wide, galvanized	"	1.20	0.47	1.67
2500	Joint clips for lath	EA.	0.24	0.24	0.48
2580	Metal base, galvanized, 2-1/2" high	L.F.	1.60	0.72	2.32
2600	Stud clips for gypsum lath	EA.	0.24	0.24	0.48
2700	Tie wire galvanized, 18 ga., 25 lb. hank	"			47.00
8000	Sound deadening board, 1/4", nailed or clipped	S.F.	0.80	0.45	1.25
09210.10	**PLASTER**				
0980	Gypsum plaster, trowel finish, 2 coats				
1000	Ceilings	S.Y.	14.75	6.20	20.95
1020	Walls	"	14.00	6.20	20.20
1030	3 coats				
1040	Ceilings	S.Y.	20.75	8.60	29.35
1060	Walls	"	18.25	8.60	26.85
7000	On columns, add to installation, 50%	"			
7020	Chases, fascia, and soffits, add to installation, 50%	"			
7040	Beams, add to installation, 50%	"			
09220.10	**PORTLAND CEMENT PLASTER**				
2980	Stucco, portland, gray, 3 coat, 1" thick				
3000	Sand finish	S.Y.	20.75	7.75	28.50
3020	Trowel finish	"	21.50	7.75	29.25
3030	White cement				
3040	Sand finish	S.Y.	21.50	8.85	30.35
3060	Trowel finish	"	23.75	8.85	32.60
3980	Scratch coat				
4000	For ceramic tile	S.Y.	4.75	2.81	7.56
4020	For quarry tile	"	4.75	2.81	7.56
5000	Portland cement plaster				
5020	2 coats, 1/2"	S.Y.	9.51	5.58	15.09
5040	3 coats, 7/8"	"	12.00	6.66	18.66

DIVISION # 09 FINISHES

		UNIT	LABOR	MAT.	TOTAL
09250.10	**GYPSUM BOARD**				
0220	1/2", clipped to				
0240	Metal furred ceiling	S.F.	0.53	0.36	0.89
0260	Columns and beams	"	1.20	0.36	1.56
0270	Walls	"	0.48	0.36	0.84
0280	Nailed or screwed to				
0290	Wood framed ceiling	S.F.	0.48	0.36	0.84
0300	Columns and beams	"	1.07	0.36	1.43
0400	Walls	"	0.43	0.36	0.79
1000	5/8", clipped to				
1020	Metal furred ceiling	S.F.	0.60	0.38	0.98
1040	Columns and beams	"	1.34	0.38	1.72
1060	Walls	"	0.53	0.38	0.91
1070	Nailed or screwed to				
1080	Wood framed ceiling	S.F.	0.60	0.38	0.98
1100	Columns and beams	"	1.34	0.38	1.72
1120	Walls	"	0.53	0.38	0.91
1122	Vinyl faced, clipped to metal studs				
1124	1/2"	S.F.	0.60	1.76	2.36
1126	5/8"	"	0.60	1.87	2.47
1130	Add for				
1140	Fire resistant	S.F.			0.10
1180	Water resistant	"			0.17
1200	Water and fire resistant	"			0.21
1220	Taping and finishing joints				
1222	Minimum	S.F.	0.32	0.04	0.36
1224	Average	"	0.40	0.06	0.46
1226	Maximum	"	0.48	0.09	0.57
5020	Casing bead				
5022	Minimum	L.F.	1.37	0.14	1.51
5024	Average	"	1.60	0.16	1.76
5026	Maximum	"	2.41	0.20	2.61
5040	Corner bead				
5042	Minimum	L.F.	1.37	0.16	1.53
5044	Average	"	1.60	0.20	1.80
5046	Maximum	"	2.41	0.25	2.66
09310.10	**CERAMIC TILE**				
0980	Glazed wall tile, 4-1/4" x 4-1/4"				
1000	Minimum	S.F.	3.33	2.32	5.65
1020	Average	"	3.89	3.64	7.53
1040	Maximum	"	4.67	9.95	14.62
2960	Base, 4-1/4" high				
2980	Minimum	L.F.	5.84	3.56	9.40
3000	Average	"	5.84	3.99	9.83
3040	Maximum	"	5.84	6.33	12.17
6100	Unglazed floor tile				
6120	Portland cement bed, cushion edge, face mounted				
6140	1" x 1"	S.F.	4.24	6.71	10.95
6160	1" x 2"	"	4.06	7.09	11.15
6180	2" x 2"	"	3.89	7.09	10.98
6200	Adhesive bed, with white grout				
6220	1" x 1"	S.F.	4.24	6.71	10.95

		UNIT	LABOR	MAT.	TOTAL
09310.10	**CERAMIC TILE, Cont'd...**				
6240	1" x 2"	S.F.	4.06	7.09	11.15
6260	2" x 2"	"	3.89	7.09	10.98
6300	Organic adhesive bed, thin set, back mounted				
6320	1" x 1"	S.F.	4.24	6.71	10.95
6340	1" x 2"	"	4.06	7.09	11.15
6350	2" x 2"	"	3.89	7.09	10.98
6360	For group 2 colors, add to material, 10%				
6370	For group 3 colors, add to material, 20%				
6380	For abrasive surface, add to material, 25%				
8990	Ceramic accessories				
9000	Towel bar, 24" long				
9004	Average	EA.	23.25	34.75	58.00
9020	Soap dish				
9024	Average	EA.	39.00	28.25	67.25
09330.10	**QUARRY TILE**				
1060	Floor				
1080	4 x 4 x 1/2"	S.F.	6.22	5.40	11.62
1120	6 x 6 x 3/4"	"	5.84	6.45	12.29
1200	Wall, applied to 3/4" portland cement bed				
1220	4 x 4 x 1/2"	S.F.	9.34	4.92	14.26
1240	6 x 6 x 3/4"	"	7.78	6.18	13.96
1320	Cove base				
1330	5 x 6 x 1/2" straight top	L.F.	7.78	4.84	12.62
1340	6 x 6 x 3/4" round top	"	7.78	5.11	12.89
1360	Stair treads 6 x 6 x 3/4"	"	11.75	7.53	19.28
1380	Window sill 6 x 8 x 3/4"	"	9.34	6.87	16.21
1400	For abrasive surface, add to material, 25%				
09410.10	**TERRAZZO**				
1100	Floors on concrete, 1-3/4" thick, 5/8" topping				
1120	Gray cement	S.F.	6.79	4.23	11.02
1140	White cement	"	6.79	4.62	11.41
1200	Sand cushion, 3" thick, 5/8" top, 1/4"				
1220	Gray cement	S.F.	7.92	5.00	12.92
1240	White cement	"	7.92	5.55	13.47
1260	Monolithic terrazzo, 3-1/2" base slab, 5/8" topping	"	5.94	3.97	9.91
1280	Terrazzo wainscot, cast-in-place, 1/2" thick	"	12.00	7.48	19.48
1300	Base, cast in place, terrazzo cove type, 6" high	L.F.	6.79	8.86	15.65
1320	Curb, cast in place, 6" wide x 6" high, polished top	"	23.75	9.84	33.59
1400	For venetian type terrazzo, add to material, 10%				
1420	For abrasive heavy duty terrazzo, add to material, 15%				
1480	Divider strips				
1500	Zinc	L.F.			1.50
1510	Brass	"			2.80
09510.10	**CEILINGS AND WALLS**				
1520	Acoustical panels, suspension system not included				
1540	Fiberglass panels				
1550	5/8" thick				
1560	2' x 2'	S.F.	0.68	0.93	1.61
1580	2' x 4'	"	0.53	0.93	1.46
1590	3/4" thick				
1600	2' x 2'	S.F.	0.68	1.54	2.22

		UNIT	LABOR	MAT.	TOTAL
09510.10	**CEILINGS AND WALLS, Cont'd...**				
1620	2' x 4'	S.F.	0.53	1.54	2.07
1640	Glass cloth faced fiberglass panels				
1660	3/4" thick	S.F.	0.80	2.55	3.35
1680	1" thick	"	0.80	2.83	3.63
1700	Mineral fiber panels				
1710	5/8" thick				
1720	2' x 2'	S.F.	0.68	1.12	1.80
1740	2' x 4'	"	0.53	1.12	1.65
1750	3/4" thick				
1760	2' x 2'	S.F.	0.68	1.51	2.19
1780	2' x 4'	"	0.53	1.51	2.04
3000	Acoustical tiles, suspension system not included				
3020	Fiberglass tile, 12" x 12"				
3040	5/8" thick	S.F.	0.87	0.71	1.58
3060	3/4" thick	"	1.07	0.93	2.00
3080	Glass cloth faced fiberglass tile				
3100	3/4" thick	S.F.	1.07	2.42	3.49
3120	3" thick	"	1.20	2.69	3.89
3140	Mineral fiber tile, 12" x 12"				
3150	5/8" thick				
3160	Standard	S.F.	0.96	0.66	1.62
3180	Vinyl faced	"	0.96	1.65	2.61
3190	3/4" thick				
3195	Standard	S.F.	0.96	1.21	2.17
3200	Vinyl faced	"	0.96	2.11	3.07
5500	Ceiling suspension systems				
5505	T bar system				
5510	2' x 4'	S.F.	0.48	1.15	1.63
5520	2' x 2'	"	0.53	1.25	1.78
5530	Concealed Z bar suspension system, 12" module	"	0.80	1.07	1.87
5550	For 1-1/2" carrier channels, 4' o.c., add	"			0.38
5560	Carrier channel for recessed light fixtures	"			0.69
09550.10	**WOOD FLOORING**				
0100	Wood strip flooring, unfinished				
1000	Fir floor				
1010	C and better				
1020	Vertical grain	S.F.	1.60	3.52	5.12
1040	Flat grain	"	1.60	3.32	4.92
1060	Oak floor				
1080	Minimum	S.F.	2.29	3.98	6.27
1100	Average	"	2.29	4.71	7.00
1120	Maximum	"	2.29	5.94	8.23
1340	Added costs				
1350	For factory finish, add to material, 10%				
1355	For random width floor, add to total, 20%				
1360	For simulated pegs, add to total, 10%				
3000	Gym floor, 2 ply felt, 25/32" maple, finished, in mastic	S.F.	2.68	7.75	10.43
3020	Over wood sleepers	"	3.01	8.69	11.70
9020	Finishing, sand, fill, finish, and wax	"	1.20	0.66	1.86
9100	Refinish sand, seal, and 2 coats of polyurethane	"	1.60	1.15	2.75
9540	Clean and wax floors	"	0.24	0.19	0.43

DIVISION # 09 FINISHES

		UNIT	LABOR	MAT.	TOTAL
09550.20	**BAMBOO FLOORING**				
0010	Vertical, Carbonized Medium, 3' vertical grain	S.F.	1.60	4.95	6.55
0020	Natural	"	1.60	4.40	6.00
0030	3' horizontal grain	"	1.60	4.95	6.55
0040	Natural	"	1.60	4.40	6.00
0050	3' Stained	"	1.60	5.50	7.10
0060	6' spice	"	1.60	7.70	9.30
0070	3' stained, butterscotch	"	1.60	5.50	7.10
0080	6' tiger	"	1.60	7.70	9.30
0090	3' stained, Irish moss	"	1.60	5.50	7.10
0100	Vice-lock, 12 mm., laminate flooring, maple	"	1.60	3.85	5.45
0101	Oak	"	1.60	3.30	4.90
0102	Pine	"	1.60	4.40	6.00
0103	Espresso	"	1.60	4.40	6.00
0104	Standard, hard maple	"	1.60	3.57	5.17
0105	Cherry	"	1.60	3.30	4.90
0106	Oak	"	1.60	2.20	3.80
0107	Walnut	"	1.60	2.20	3.80
0108	South pacific vice-lock, 12 mm, brazilian cherry	"	1.60	4.95	6.55
0109	Maple	"	1.60	4.40	6.00
0110	Teak	"	1.60	4.67	6.27
09630.10	**UNIT MASONRY FLOORING**				
1000	Clay brick				
1020	9 x 4-1/2 x 3" thick				
1040	Glazed	S.F.	4.02	8.03	12.05
1060	Unglazed	"	4.02	7.70	11.72
1070	8 x 4 x 3/4" thick				
1080	Glazed	S.F.	4.19	7.26	11.45
1100	Unglazed	"	4.19	7.15	11.34
1140	For herringbone pattern, add to labor, 15%				
09660.10	**RESILIENT TILE FLOORING**				
1020	Solid vinyl tile, 1/8" thick, 12" x 12"				
1040	Marble patterns	S.F.	1.20	4.45	5.65
1060	Solid colors	"	1.20	5.77	6.97
1080	Travertine patterns	"	1.20	6.49	7.69
2000	Conductive resilient flooring, vinyl tile				
2040	1/8" thick, 12" x 12"	S.F.	1.37	6.71	8.08
09665.10	**RESILIENT SHEET FLOORING**				
0980	Vinyl sheet flooring				
1000	Minimum	S.F.	0.48	2.86	3.34
1002	Average	"	0.58	4.62	5.20
1004	Maximum	"	0.80	7.81	8.61
1020	Cove, to 6"	L.F.	0.96	1.70	2.66
2000	Fluid applied resilient flooring				
2020	Polyurethane, poured in place, 3/8" thick	S.F.	4.02	10.50	14.52
6200	Vinyl sheet goods, backed				
6220	0.070" thick	S.F.	0.60	3.90	4.50
6260	0.125" thick	"	0.60	6.98	7.58
6280	0.250" thick	"	0.60	8.03	8.63

		UNIT	LABOR	MAT.	TOTAL
09678.10	**RESILIENT BASE AND ACCESSORIES**				
1000	Wall base, vinyl				
1130	4" high	L.F.	1.60	1.04	2.64
1140	6" high	"	1.60	1.43	3.03
09682.10	**CARPET PADDING**				
1000	Carpet padding				
1005	Foam rubber, waffle type, 0.3" thick	S.Y.	2.41	6.16	8.57
1010	Jute padding				
1022	Average	S.Y.	2.41	5.44	7.85
1030	Sponge rubber cushion				
1042	Average	S.Y.	2.41	6.60	9.01
1050	Urethane cushion, 3/8" thick				
1062	Average	S.Y.	2.41	5.77	8.18
09685.10	**CARPET**				
0990	Carpet, acrylic				
1000	24 oz., light traffic	S.Y.	5.36	28.00	33.36
1020	28 oz., medium traffic	"	5.36	36.25	41.61
2010	Nylon				
2020	15 oz., light traffic	S.Y.	5.36	20.25	25.61
2040	28 oz., medium traffic	"	5.36	26.50	31.86
2110	Nylon				
2120	28 oz., medium traffic	S.Y.	5.36	25.25	30.61
2140	35 oz., heavy traffic	"	5.36	30.75	36.11
2145	Wool				
2150	30 oz., medium traffic	S.Y.	5.36	41.75	47.11
2160	36 oz., medium traffic	"	5.36	44.00	49.36
2180	42 oz., heavy traffic	"	5.36	58.00	63.36
3000	Carpet tile				
3020	Foam backed				
3022	Minimum	S.F.	0.96	3.21	4.17
3024	Average	"	1.07	3.71	4.78
3026	Maximum	"	1.20	5.88	7.08
8980	Clean and vacuum carpet				
9000	Minimum	S.Y.	0.18	0.29	0.47
9020	Average	"	0.32	0.46	0.78
9040	Maximum	"	0.48	0.63	1.11
09905.10	**PAINTING PREPARATION**				
1000	Dropcloths				
1050	Minimum	S.F.	0.03	0.02	0.05
1100	Average	"	0.04	0.03	0.07
1150	Maximum	"	0.05	0.04	0.09
1200	Masking				
1250	Paper and tape				
1300	Minimum	L.F.	0.48	0.02	0.50
1350	Average	"	0.60	0.03	0.63
1400	Maximum	"	0.80	0.04	0.84
1450	Doors				
1500	Minimum	EA.	6.03	0.04	6.07
1550	Average	"	8.04	0.05	8.09
1600	Maximum	"	10.75	0.06	10.81
1650	Windows				
1700	Minimum	EA.	6.03	0.04	6.07

DIVISION # 09 FINISHES

		UNIT	LABOR	MAT.	TOTAL
09905.10	**PAINTING PREPARATION, Cont'd...**				
1750	Average	EA.	8.04	0.05	8.09
1800	Maximum	"	10.75	0.06	10.81
2000	Sanding				
2050	Walls and flat surfaces				
2100	Minimum	S.F.	0.32		0.32
2150	Average	"	0.40		0.40
2200	Maximum	"	0.48		0.48
2250	Doors and windows				
2300	Minimum	EA.	8.04		8.04
2350	Average	"	12.00		12.00
2400	Maximum	"	16.00		16.00
2450	Trim				
2500	Minimum	L.F.	0.60		0.60
2550	Average	"	0.80		0.80
2600	Maximum	"	1.07		1.07
2650	Puttying				
2700	Minimum	S.F.	0.74	0.01	0.75
2750	Average	"	0.96	0.02	0.98
2800	Maximum	"	1.20	0.03	1.23
09910.05	**EXT. PAINTING, SITEWORK**				
3000	Concrete Block				
3020	Roller				
3040	First Coat				
3060	Minimum	S.F.	0.24	0.22	0.46
3080	Average	"	0.32	0.22	0.54
3100	Maximum	"	0.48	0.22	0.70
3120	Second Coat				
3140	Minimum	S.F.	0.20	0.22	0.42
3160	Average	"	0.26	0.22	0.48
3180	Maximum	"	0.40	0.22	0.62
3200	Spray				
3220	First Coat				
3240	Minimum	S.F.	0.13	0.17	0.30
3260	Average	"	0.16	0.17	0.33
3280	Maximum	"	0.18	0.17	0.35
3300	Second Coat				
3320	Minimum	S.F.	0.08	0.17	0.25
3340	Average	"	0.10	0.17	0.27
3360	Maximum	"	0.15	0.17	0.32
3500	Fences, Chain Link				
3700	Roller				
3720	First Coat				
3740	Minimum	S.F.	0.34	0.13	0.47
3760	Average	"	0.40	0.13	0.53
3780	Maximum	"	0.45	0.13	0.58
3800	Second Coat				
3820	Minimum	S.F.	0.20	0.13	0.33
3840	Average	"	0.24	0.13	0.37
3860	Maximum	"	0.30	0.13	0.43
3880	Spray				
3900	First Coat				

		UNIT	LABOR	MAT.	TOTAL
09910.05	**EXT. PAINTING, SITEWORK, Cont'd...**				
3920	Minimum	S.F.	0.15	0.11	0.26
3940	Average	"	0.17	0.11	0.28
3960	Maximum	"	0.20	0.11	0.31
3980	Second Coat				
4000	Minimum	S.F.	0.11	0.11	0.22
4060	Average	"	0.13	0.11	0.24
4080	Maximum	"	0.15	0.11	0.26
4200	Fences, Wood or Masonry				
4220	Brush				
4240	First Coat				
4260	Minimum	S.F.	0.50	0.22	0.72
4280	Average	"	0.60	0.22	0.82
4300	Maximum	"	0.80	0.22	1.02
4320	Second Coat				
4340	Minimum	S.F.	0.30	0.22	0.52
4360	Average	"	0.37	0.22	0.59
4380	Maximum	"	0.48	0.22	0.70
4400	Roller				
4420	First Coat				
4440	Minimum	S.F.	0.26	0.22	0.48
4460	Average	"	0.32	0.22	0.54
4480	Maximum	"	0.37	0.22	0.59
4500	Second Coat				
4520	Minimum	S.F.	0.18	0.22	0.40
4540	Average	"	0.22	0.22	0.44
4560	Maximum	"	0.30	0.22	0.52
4580	Spray				
4600	First Coat				
4620	Minimum	S.F.	0.17	0.17	0.34
4640	Average	"	0.21	0.17	0.38
4660	Maximum	"	0.30	0.17	0.47
4680	Second Coat				
4700	Minimum	S.F.	0.12	0.17	0.29
4760	Average	"	0.15	0.17	0.32
4780	Maximum	"	0.20	0.17	0.37
09910.15	**EXT. PAINTING, BUILDINGS**				
1200	Decks, Wood, Stained				
1580	Spray				
1600	First Coat				
1620	Minimum	S.F.	0.15	0.14	0.29
1640	Average	"	0.16	0.14	0.30
1660	Maximum	"	0.18	0.14	0.32
1680	Second Coat				
1700	Minimum	S.F.	0.13	0.13	0.26
1720	Average	"	0.14	0.13	0.27
1740	Maximum	"	0.16	0.13	0.29
2520	Doors, Wood				
2540	Brush				
2560	First Coat				
2580	Minimum	S.F.	0.74	0.17	0.91
2600	Average	"	0.96	0.17	1.13

		UNIT	LABOR	MAT.	TOTAL
09910.15	**EXT. PAINTING, BUILDINGS, Cont'd...**				
2620	Maximum	S.F.	1.20	0.17	1.37
2640	Second Coat				
2660	Minimum	S.F.	0.60	0.17	0.77
2680	Average	"	0.68	0.17	0.85
2700	Maximum	"	0.80	0.17	0.97
3680	Siding, Wood				
3880	Spray				
3900	First Coat				
3920	Minimum	S.F.	0.16	0.13	0.29
3940	Average	"	0.17	0.13	0.30
3960	Maximum	"	0.18	0.13	0.31
3980	Second Coat				
4000	Minimum	S.F.	0.12	0.13	0.25
4020	Average	"	0.16	0.13	0.29
4040	Maximum	"	0.24	0.13	0.37
4440	Trim				
4460	Brush				
4480	First Coat				
4500	Minimum	L.F.	0.20	0.22	0.42
4520	Average	"	0.24	0.22	0.46
4540	Maximum	"	0.30	0.22	0.52
4560	Second Coat				
4580	Minimum	L.F.	0.15	0.22	0.37
4600	Average	"	0.20	0.22	0.42
4620	Maximum	"	0.30	0.22	0.52
4640	Walls				
4840	Spray				
4860	First Coat				
4880	Minimum	S.F.	0.07	0.13	0.20
4900	Average	"	0.09	0.13	0.22
4920	Maximum	"	0.12	0.13	0.25
4940	Second Coat				
4960	Minimum	S.F.	0.06	0.13	0.19
4980	Average	"	0.08	0.13	0.21
5000	Maximum	"	0.10	0.13	0.23
5020	Windows				
5040	Brush				
5060	First Coat				
5080	Minimum	S.F.	0.80	0.15	0.95
5100	Average	"	0.96	0.15	1.11
5120	Maximum	"	1.20	0.15	1.35
5140	Second Coat				
5160	Minimum	S.F.	0.68	0.15	0.83
5180	Average	"	0.80	0.15	0.95
5200	Maximum	"	0.96	0.15	1.11
09910.35	**INT. PAINTING, BUILDINGS**				
1380	Cabinets and Casework				
1400	Brush				
1420	First Coat				
1440	Minimum	S.F.	0.48	0.22	0.70
1460	Average	"	0.53	0.22	0.75

DIVISION # 09 FINISHES

		UNIT	LABOR	MAT.	TOTAL
09910.35	**INT. PAINTING, BUILDINGS, Cont'd...**				
1480	Maximum	S.F.	0.60	0.22	0.82
1500	Second Coat				
1520	Minimum	S.F.	0.40	0.22	0.62
1540	Average	"	0.43	0.22	0.65
1560	Maximum	"	0.48	0.22	0.70
1580	Spray				
1600	First Coat				
1620	Minimum	S.F.	0.24	0.17	0.41
1640	Average	"	0.28	0.17	0.45
1660	Maximum	"	0.34	0.17	0.51
1680	Second Coat				
1700	Minimum	S.F.	0.19	0.17	0.36
1720	Average	"	0.20	0.17	0.37
1740	Maximum	"	0.26	0.17	0.43
2520	Doors, Wood				
2540	Brush				
2560	First Coat				
2580	Minimum	S.F.	0.68	0.22	0.90
2600	Average	"	0.87	0.22	1.09
2620	Maximum	"	1.07	0.22	1.29
2640	Second Coat				
2660	Minimum	S.F.	0.53	0.16	0.69
2680	Average	"	0.60	0.16	0.76
2700	Maximum	"	0.68	0.16	0.84
2720	Spray				
2740	First Coat				
2760	Minimum	S.F.	0.14	0.15	0.29
2780	Average	"	0.17	0.15	0.32
2800	Maximum	"	0.21	0.15	0.36
2820	Second Coat				
2840	Minimum	S.F.	0.11	0.15	0.26
2860	Average	"	0.13	0.15	0.28
2880	Maximum	"	0.15	0.15	0.30
3900	Trim				
3920	Brush				
3940	First Coat				
3960	Minimum	L.F.	0.19	0.22	0.41
3980	Average	"	0.21	0.22	0.43
4000	Maximum	"	0.26	0.22	0.48
4020	Second Coat				
4040	Minimum	L.F.	0.14	0.22	0.36
4060	Average	"	0.18	0.22	0.40
4080	Maximum	"	0.26	0.22	0.48
4100	Walls				
4120	Roller				
4140	First Coat				
4160	Minimum	S.F.	0.17	0.17	0.34
4180	Average	"	0.17	0.17	0.34
4200	Maximum	"	0.20	0.17	0.37
4220	Second Coat				
4240	Minimum	S.F.	0.15	0.17	0.32
4260	Average	"	0.16	0.17	0.33

		UNIT	LABOR	MAT.	TOTAL
09910.35	**INT. PAINTING, BUILDINGS, Cont'd...**				
4280	Maximum	S.F.	0.18	0.17	0.35
4300	Spray				
4320	First Coat				
4340	Minimum	S.F.	0.07	0.14	0.21
4360	Average	"	0.09	0.14	0.23
4380	Maximum	"	0.12	0.14	0.26
4400	Second Coat				
4420	Minimum	S.F.	0.06	0.14	0.20
4440	Average	"	0.08	0.14	0.22
4460	Maximum	"	0.10	0.14	0.24
09955.10	**WALL COVERING**				
0900	Vinyl wall covering				
1000	Medium duty	S.F.	0.68	1.43	2.11
1010	Heavy duty	"	0.80	2.09	2.89

Design & Construction Resources

TABLE OF CONTENTS   PAGE

10110.10 - CHALKBOARDS	**10-2**
10165.10 - TOILET PARTITIONS	**10-2**
10185.10 - SHOWER STALLS	**10-2**
10290.10 - PEST CONTROL	**10-3**
10350.10 - FLAGPOLES	**10-3**
10400.10 - IDENTIFYING DEVICES	**10-3**
10450.10 - CONTROL	**10-5**
10500.10 - LOCKERS	**10-5**
10520.10 - FIRE PROTECTION	**10-5**
10550.10 - POSTAL SPECIALTIES	**10-6**
10670.10 - SHELVING	**10-6**
10800.10 - BATH ACCESSORIES	**10-6**

DIVISION # 10 SPECIALTIES

		UNIT	LABOR	MAT.	TOTAL
10110.10	**CHALKBOARDS**				
1020	Chalkboard, metal frame, 1/4" thick				
1040	48"x60"	EA.	48.25	430	478
1060	48"x96"	"	54.00	650	704
1080	48"x144"	"	60.00	720	780
1100	48"x192"	"	69.00	960	1,029
1110	Liquid chalkboard				
1120	48"x60"	EA.	48.25	380	428
1140	48"x96"	"	54.00	480	534
1160	48"x144"	"	60.00	600	660
1180	48"x192"	"	69.00	690	759
1200	Map rail, deluxe	L.F.	2.41	8.80	11.21
1388	Average	PCT.			6.00
10165.10	**TOILET PARTITIONS**				
0100	Toilet partition, plastic laminate				
0120	Ceiling mounted	EA.	160	930	1,090
0140	Floor mounted	"	120	890	1,010
0150	Metal				
0165	Ceiling mounted	EA.	160	740	900
0180	Floor mounted	"	120	700	820
0190	Wheel chair partition, plastic laminate				
0200	Ceiling mounted	EA.	160	430	590
0210	Floor mounted	"	120	290	410
0215	Painted metal				
0220	Ceiling mounted	EA.	160	390	550
0230	Floor mounted	"	120	280	400
1980	Urinal screen, plastic laminate				
2000	Wall hung	EA.	60.00	480	540
2100	Floor mounted	"	60.00	400	460
2120	Porcelain enameled steel, floor mounted	"	60.00	560	620
2140	Painted metal, floor mounted	"	60.00	360	420
2160	Stainless steel, floor mounted	"	60.00	940	1,000
5000	Metal toilet partitions				
5020	Front door and side divider, floor mounted				
5040	Porcelain enameled steel	EA.	120	1,260	1,380
5060	Painted steel	"	120	420	540
5080	Stainless steel	"	120	1,280	1,400
10185.10	**SHOWER STALLS**				
1000	Shower receptors				
1010	Precast, terrazzo				
1020	32" x 32"	EA.	44.50	470	515
1040	32" x 48"	"	53.00	530	583
1050	Concrete				
1060	32" x 32"	EA.	44.50	280	325
1080	48" x 48"	"	59.00	330	389
1100	Shower door, trim and hardware				
1130	Porcelain enameled steel, flush	EA.	53.00	530	583
1140	Baked enameled steel, flush	"	53.00	370	423
1150	Aluminum frame, tempered glass, 48" wide, sliding	"	67.00	570	637
1161	Folding	"	67.00	540	607
5400	Shower compartment, precast concrete receptor				
5420	Single entry type				

		UNIT	LABOR	MAT.	TOTAL
10185.10	**SHOWER STALLS, Cont'd...**				
5440	Porcelain enameled steel	EA.	530	2,620	3,150
5460	Baked enameled steel	"	530	1,540	2,070
5480	Stainless steel	"	530	3,110	3,640
5500	Double entry type				
5520	Porcelain enameled steel	EA.	670	5,490	6,160
5540	Baked enameled steel	"	670	1,520	2,190
5560	Stainless steel	"	670	2,970	3,640
10290.10	**PEST CONTROL**				
1000	Termite control				
1010	Under slab spraying				
1020	Minimum	S.F.	0.09	0.41	0.50
1040	Average	"	0.18	0.55	0.73
1120	Maximum	"	0.37	0.82	1.19
10350.10	**FLAGPOLES**				
2020	Installed in concrete base				
2030	Fiberglass				
2040	25' high	EA.	320	1,460	1,780
2080	50' high	"	800	6,480	7,280
2100	Aluminum				
2120	25' high	EA.	320	2,120	2,440
2140	50' high	"	800	6,930	7,730
2160	Bonderized steel				
2180	25' high	EA.	370	3,520	3,890
2200	50' high	"	970	8,580	9,550
2220	Freestanding tapered, fiberglass				
2240	30' high	EA.	340	1,460	1,800
2260	40' high	"	440	2,020	2,460
2280	50' high	"	480	6,480	6,960
2300	60' high	"	570	7,770	8,340
2400	Wall mounted, with collar, brushed aluminum finish				
2420	15' long	EA.	240	1,210	1,450
2440	18' long	"	240	1,320	1,560
2460	20' long	"	250	1,480	1,730
2480	24' long	"	280	1,590	1,870
2500	Outrigger, wall, including base				
2520	10' long	EA.	320	680	1,000
2540	20' long	"	400	1,230	1,630
10400.10	**IDENTIFYING DEVICES**				
1000	Directory and bulletin boards				
1020	Open face boards				
1040	Chrome plated steel frame	S.F.	24.25	27.50	51.75
1060	Aluminum framed	"	24.25	23.00	47.25
1080	Bronze framed	"	24.25	30.75	55.00
1100	Stainless steel framed	"	24.25	55.00	79.25
1140	Tack board, aluminum framed	"	24.25	38.50	62.75
1160	Visual aid board, aluminum framed	"	24.25	110	134
1200	Glass encased boards, hinged and keyed				
1210	Aluminum framed	S.F.	60.00	120	180
1220	Bronze framed	"	60.00	120	180
1230	Stainless steel framed	"	60.00	210	270
1240	Chrome plated steel framed	"	60.00	200	260

		UNIT	LABOR	MAT.	TOTAL
10400.10	**IDENTIFYING DEVICES, Cont'd...**				
2020	Metal plaque				
2040	Cast bronze	S.F.	40.25	240	280
2060	Aluminum	"	40.25	190	230
2080	Metal engraved plaque				
2100	Porcelain steel	S.F.	40.25	210	250
2120	Stainless steel	"	40.25	190	230
2140	Brass	"	40.25	150	190
2160	Aluminum	"	40.25	130	170
2200	Metal built-up plaque				
2220	Bronze	S.F.	48.25	160	208
2240	Copper and bronze	"	48.25	140	188
2260	Copper and aluminum	"	48.25	120	168
2280	Metal nameplate plaques				
2300	Cast bronze	S.F.	30.25	1,730	1,760
2320	Cast aluminum	"	30.25	1,110	1,140
2330	Engraved, 1-1/2" x 6"				
2340	Bronze	EA.	30.25	200	230
2360	Aluminum	"	30.25	200	230
2440	Letters, on masonry or concrete, aluminum, satin finish				
2450	1/2" thick				
2460	2" high	EA.	19.25	19.75	39.00
2480	4" high	"	24.25	29.75	54.00
2500	6" high	"	26.75	39.75	66.50
2510	3/4" thick				
2520	8" high	EA.	30.25	60.00	90.25
2540	10" high	"	34.50	69.00	104
2550	1" thick				
2560	12" high	EA.	40.25	99.00	139
2580	14" high	"	48.25	120	168
2600	16" high	"	60.00	150	210
2620	For polished aluminum add, 15%				
2640	For clear anodized aluminum add, 15%				
2660	For colored anodic aluminum add, 30%				
2680	For profiled and color enameled letters add, 50%				
2700	Cast bronze, satin finish letters				
2710	3/8" thick				
2720	2" high	EA.	19.25	52.00	71.25
2740	4" high	"	24.25	67.00	91.25
2760	1/2" thick, 6" high	"	26.75	91.00	118
2780	5/8" thick, 8" high	"	30.25	120	150
2785	1" thick				
2790	10" high	EA.	34.50	180	215
2800	12" high	"	40.25	210	250
2820	14" high	"	48.25	270	318
2840	16" high	"	60.00	350	410
3000	Interior door signs, adhesive, flexible				
3060	2" x 8"	EA.	9.38	35.75	45.13
3080	4" x 4"	"	9.38	38.50	47.88
3100	6" x 7"	"	9.38	42.25	51.63
3120	6" x 9"	"	9.38	42.50	51.88
3140	10" x 9"	"	9.38	49.50	58.88
3160	10" x 12"	"	9.38	61.00	70.38

DIVISION # 10 SPECIALTIES

		UNIT	LABOR	MAT.	TOTAL
10400.10	**IDENTIFYING DEVICES, Cont'd...**				
3200	Hard plastic type, no frame				
3220	3" x 8"	EA.	9.38	8.80	18.18
3240	4" x 4"	"	9.38	13.25	22.63
3260	4" x 12"	"	9.38	25.25	34.63
3280	Hard plastic type, with frame				
3300	3" x 8"	EA.	9.38	14.25	23.63
3320	4" x 4"	"	9.38	20.75	30.13
3340	4" x 12"	"	9.38	27.50	36.88
10450.10	**CONTROL**				
1020	Access control, 7' high, indoor or outdoor impenetrability				
1040	Remote or card control, type B	EA.	660	4,390	5,050
1060	Free passage, type B	"	660	3,980	4,640
1080	Remote or card control, type AA	"	660	5,720	6,380
1100	Free passage, type AA	"	660	5,390	6,050
10500.10	**LOCKERS**				
0080	Locker bench, floor mounted, laminated maple				
0100	4'	EA.	40.25	220	260
0120	6'	"	40.25	320	360
0130	Wardrobe locker, 12" x 60" x 15", baked on enamel				
0140	1-tier	EA.	24.25	240	264
0160	2-tier	"	24.25	260	284
0180	3-tier	"	25.50	260	286
0200	4-tier	"	25.50	320	346
0240	12" x 72" x 15", baked on enamel				
0260	1-tier	EA.	24.25	260	284
0280	2-tier	"	24.25	270	294
0300	4-tier	"	25.50	310	336
0320	5-tier	"	25.50	310	336
1200	15" x 60" x 15", baked on enamel				
1220	1-tier	EA.	24.25	260	284
1240	4-tier	"	25.50	290	316
10520.10	**FIRE PROTECTION**				
1000	Portable fire extinguishers				
1020	Water pump tank type				
1030	2.5 gal.				
1040	Red enameled galvanized	EA.	25.00	130	155
1060	Red enameled copper	"	25.00	140	165
1080	Polished copper	"	25.00	240	265
1200	Carbon dioxide type, red enamel steel				
1210	Squeeze grip with hose and horn				
1220	2.5 lb	EA.	25.00	160	185
1240	5 lb	"	28.75	220	249
1260	10 lb	"	37.50	300	338
1280	15 lb	"	47.00	350	397
1300	20 lb	"	47.00	420	467
1310	Wheeled type				
1320	125 lb	EA.	75.00	2,200	2,275
1340	250 lb	"	75.00	3,850	3,925
1360	500 lb	"	75.00	5,720	5,795
1400	Dry chemical, pressurized type				
1405	Red enameled steel				

		UNIT	LABOR	MAT.	TOTAL
10520.10	**FIRE PROTECTION, Cont'd...**				
1410	2.5 lb	EA.	25.00	22.00	47.00
1430	5 lb	"	28.75	44.00	72.75
1440	10 lb	"	37.50	66.00	104
1450	20 lb	"	47.00	120	167
1460	30 lb	"	47.00	160	207
1480	Chrome plated steel, 2.5 lb	"	25.00	220	245
10550.10	**POSTAL SPECIALTIES**				
1500	Mail chutes				
1520	Single mail chute				
1530	Finished aluminum	L.F.	120	1,290	1,410
1540	Bronze	"	120	1,840	1,960
1560	Single mail chute receiving box				
1580	Finished aluminum	EA.	240	1,240	1,480
1600	Bronze	"	240	2,040	2,280
1620	Twin mail chute, double parallel				
1630	Finished aluminum	FLR	240	1,810	2,050
1640	Bronze	"	240	2,200	2,440
10670.10	**SHELVING**				
0980	Shelving, enamel, closed side and back, 12" x 36"				
1000	5 shelves	EA.	80.00	230	310
1020	8 shelves	"	110	260	370
1030	Open				
1040	5 shelves	EA.	80.00	150	230
1060	8 shelves	"	110	170	280
2000	Metal storage shelving, baked enamel				
2030	7 shelf unit, 72" or 84" high				
2050	12" shelf	L.F.	51.00	63.00	114
2080	24" shelf	"	60.00	83.00	143
2100	36" shelf	"	69.00	110	179
2200	4 shelf unit, 40" high				
2240	12" shelf	L.F.	44.00	52.00	96.00
2270	24" shelf	"	54.00	76.00	130
2300	3 shelf unit, 32" high				
2340	12" shelf	L.F.	25.50	42.50	68.00
2370	24" shelf	"	30.25	53.00	83.25
2400	Single shelf unit, attached to masonry				
2420	12" shelf	L.F.	8.77	17.00	25.77
2450	24" shelf	"	10.50	21.75	32.25
2460	For stainless steel, add to material, 120%				
2470	For attachment to gypsum board, add to labor, 50%				
10800.10	**BATH ACCESSORIES**				
1040	Ash receiver, wall mounted, aluminum	EA.	24.25	99.00	123
1050	Grab bar, 1-1/2" dia., stainless steel, wall mounted				
1060	24" long	EA.	24.25	33.00	57.25
1080	36" long	"	25.50	55.00	80.50
1100	48" long	"	28.50	88.00	117
1130	1" dia., stainless steel				
1140	12" long	EA.	21.00	22.50	43.50
1180	24" long	"	24.25	33.00	57.25
1220	36" long	"	26.75	46.25	73.00
1240	48" long	"	28.50	53.00	81.50

		UNIT	LABOR	MAT.	TOTAL
10800.10	**BATH ACCESSORIES, Cont'd...**				
1300	Hand dryer, surface mounted, 110 volt	EA.	60.00	360	420
1320	Medicine cabinet, 16 x 22, baked enamel, steel, lighted	"	19.25	89.00	108
1340	With mirror, lighted	"	32.25	150	182
1420	Mirror, 1/4" plate glass, up to 10 sf	S.F.	4.82	8.96	13.78
1430	Mirror, stainless steel frame				
1440	18"x24"	EA.	16.00	58.00	74.00
1500	24"x30"	"	24.25	74.00	98.25
1520	24"x48"	"	40.25	160	200
1560	30"x30"	"	48.25	200	248
1600	48"x72"	"	80.00	450	530
1640	With shelf, 18"x24"	"	19.25	200	219
1820	Sanitary napkin dispenser, stainless steel, wall mounted	"	32.25	510	542
1830	Shower rod, 1" diameter				
1840	Chrome finish over brass	EA.	24.25	34.00	58.25
1860	Stainless steel	"	24.25	77.00	101
1900	Soap dish, stainless steel, wall mounted	"	32.25	220	252
1910	Toilet tissue dispenser, stainless, wall mounted				
1920	Single roll	EA.	12.00	170	182
1940	Double roll	"	13.75	260	274
1945	Towel dispenser, stainless steel				
1950	Flush mounted	EA.	26.75	170	197
1960	Surface mounted	"	24.25	72.00	96.25
1970	Combination towel dispenser and waste receptacle	"	32.25	720	752
2000	Towel bar, stainless steel				
2020	18" long	EA.	19.25	66.00	85.25
2040	24" long	"	22.00	74.00	96.00
2060	30" long	"	24.25	92.00	116
2070	36" long	"	26.75	78.00	105
2100	Waste receptacle, stainless steel, wall mounted	"	40.25	92.00	132

Design & Construction Resources

TABLE OF CONTENTS   PAGE

11010.10 - MAINTENANCE EQUIPMENT	**11-2**
11060.10 - THEATER EQUIPMENT	**11-2**
11090.10 - CHECKROOM EQUIPMENT	**11-2**
11110.10 - LAUNDRY EQUIPMENT	**11-2**
11161.10 - LOADING DOCK EQUIPMENT	**11-2**
11170.20 - AERATION EQUIPMENT	**11-3**
11400.10 - FOOD SERVICE EQUIPMENT	**11-3**
11450.10 - RESIDENTIAL EQUIPMENT	**11-5**
11600.10 - LABORATORY EQUIPMENT	**11-6**
11700.10 - MEDICAL EQUIPMENT	**11-6**

		UNIT	LABOR	MAT.	TOTAL
11010.10	**MAINTENANCE EQUIPMENT**				
1000	Vacuum cleaning system				
1010	3 valves				
1020	1.5 hp	EA.	540	2,850	3,390
1030	2.5 hp	"	690	3,290	3,980
1040	5 valves	"	970	3,840	4,810
1060	7 valves	"	1,210	4,390	5,600
11060.10	**THEATER EQUIPMENT**				
1000	Roll out stage, steel frame, wood floor				
1020	Manual	S.F.	3.01	40.25	43.26
1040	Electric	"	4.82	43.00	47.82
1100	Portable stages				
1120	8" high	S.F.	2.41	18.25	20.66
1140	18" high	"	2.68	19.00	21.68
1160	36" high	"	2.84	22.50	25.34
1180	48" high	"	3.01	24.50	27.51
1300	Band risers				
1320	Minimum	S.F.	2.41	35.75	38.16
1340	Maximum	"	2.41	61.00	63.41
1400	Chairs for risers				
1420	Minimum	EA.	1.70	94.00	95.70
1440	Maximum	"	1.70	160	162
2000	Theatre controlls				
2010	Fade console, 48 channel	EA.	250	4,840	5,090
2020	Light control modules, 125 channels	"	490	9,080	9,570
2030	Dimmer module, stage-pin output	"	250	2,080	2,330
2040	6-Module pack, w/24 U-ground connectors	"	250	340	590
11090.10	**CHECKROOM EQUIPMENT**				
1000	Motorized checkroom equipment				
1020	No shelf system, 6'4" height				
1040	7'6" length	EA.	480	2,670	3,150
1060	14'6" length	"	480	3,020	3,500
1080	28' length	"	480	4,160	4,640
1100	One shelf, 6'8" height				
1120	7'6" length	EA.	480	3,270	3,750
1140	14'6" length	"	480	3,960	4,440
1160	28' length	"	480	5,860	6,340
11110.10	**LAUNDRY EQUIPMENT**				
1000	High capacity, heavy duty				
1020	Washer extractors				
1030	135 lb				
1040	Standard	EA.	400	31,790	32,190
1060	Pass through	"	400	70,730	71,130
1070	200 lb				
1080	Standard	EA.	400	66,200	66,600
1100	Pass through	"	400	72,820	73,220
1120	110 lb dryer	"	400	9,630	10,030
11161.10	**LOADING DOCK EQUIPMENT**				
0080	Dock leveler, 10 ton capacity				
0100	6' x 8'	EA.	480	4,250	4,730
0120	7' x 8'	"	480	4,740	5,220

		UNIT	LABOR	MAT.	TOTAL
11170.10	**WASTE HANDLING**				
1500	Industrial compactor				
1520	1 cy	EA.	550	12,730	13,280
1540	3 cy	"	710	23,760	24,470
1560	5 cy	"	990	30,330	31,320
11170.20	**AERATION EQUIPMENT**				
0010	Surface spray/Vertical pump				
0020	1 hp. Pump, 500 gpm.	EA.	270		270
0030	5 hp. Pump, 2000 gpm.	"	1,070		1,070
0040	Polycarbon, treatment container, 1,000 gallon capacity	"	2,110	4,150	6,260
0050	1,500 gallon	"	2,110	8,640	10,750
0060	2,000 gallon	"	2,640	11,670	14,310
0070	3,000 gallon	"	2,820	15,710	18,530
11400.10	**FOOD SERVICE EQUIPMENT**				
1000	Unit kitchens				
1020	30" compact kitchen				
1040	Refrigerator, with range, sink	EA.	250	1,390	1,640
1060	Sink only	"	160	770	930
1080	Range only	"	120	660	780
1100	Cabinet for upper wall section	"	71.00	460	531
1120	Stainless shield, for rear wall	"	19.75	180	200
1140	Side wall	"	19.75	86.00	106
1200	42" compact kitchen				
1220	Refrigerator with range, sink	EA.	270	1,920	2,190
1240	Sink only	"	250	1,380	1,630
1260	Cabinet for upper wall section	"	82.00	660	742
1280	Stainless shield, for rear wall	"	20.50	210	231
1290	Side wall	"	20.50	86.00	107
1600	Bake oven				
1610	Single deck				
1620	Minimum	EA.	62.00	2,510	2,572
1640	Maximum	"	120	3,120	3,240
1650	Double deck				
1660	Minimum	EA.	82.00	3,100	3,182
1680	Maximum	"	120	5,650	5,770
1685	Triple deck				
1690	Minimum	EA.	82.00	4,130	4,212
1695	Maximum	"	160	8,970	9,130
1700	Convection type oven, electric, 40" x 45" x 57"				
1720	Minimum	EA.	62.00	4,140	4,202
1740	Maximum	"	120	8,210	8,330
1800	Broiler, without oven, 69" x 26" x 39"				
1820	Minimum	EA.	62.00	4,180	4,242
1840	Maximum	"	82.00	6,940	7,022
1900	Coffee urns, 10 gallons				
1920	Minimum	EA.	160	2,290	2,450
1940	Maximum	"	250	3,280	3,530
2000	Fryer, with submerger				
2010	Single				
2020	Minimum	EA.	99.00	1,420	1,519
2040	Maximum	"	160	4,070	4,230
2050	Double				

		UNIT	LABOR	MAT.	TOTAL
11400.10	**FOOD SERVICE EQUIPMENT, Cont'd...**				
2060	Minimum	EA.	120	2,710	2,830
2080	Maximum	"	160	6,400	6,560
2100	Griddle, counter				
2110	3' long				
2120	Minimum	EA.	82.00	1,780	1,862
2140	Maximum	"	99.00	4,460	4,559
2150	5' long				
2160	Minimum	EA.	120	2,740	2,860
2180	Maximum	"	160	7,050	7,210
2200	Kettles, steam, jacketed				
2210	20 gallons				
2220	Minimum	EA.	120	3,190	3,310
2240	Maximum	"	250	6,780	7,030
2300	Range				
2310	Heavy duty, single oven, open top				
2320	Minimum	EA.	62.00	2,200	2,262
2340	Maximum	"	160	6,650	6,810
2350	Fry top				
2360	Minimum	EA.	62.00	3,180	3,242
2380	Maximum	"	160	7,220	7,380
2400	Steamers, electric				
2410	27 kw				
2420	Minimum	EA.	120	11,160	11,280
2440	Maximum	"	160	16,030	16,190
2450	18 kw				
2460	Minimum	EA.	120	8,030	8,150
2480	Maximum	"	160	14,850	15,010
2500	Dishwasher, rack type				
2520	Single tank, 190 racks/hr	EA.	250	11,220	11,470
2530	Double tank				
2540	234 racks/hr	EA.	270	20,900	21,170
2560	265 racks/hr	"	330	22,690	23,020
2580	Dishwasher, automatic 100 meals/hr	"	160	65,890	66,050
2600	Disposals				
2620	100 gal/hr	EA.	160	1,930	2,090
2640	120 gal/hr	"	170	2,480	2,650
2660	250 gal/hr	"	180	6,310	6,490
2680	Exhaust hood for dishwasher, gutter 4 sides, s-steel				
2681	4'x4'x2'	EA.	180	2,560	2,740
2690	4'x7'x2'	"	200	3,750	3,950
2900	Ice cube maker				
2910	50 lb per day				
2920	Minimum	EA.	490	1,660	2,150
2940	Maximum	"	490	3,520	4,010
2950	500 lb per day				
2960	Minimum	EA.	820	4,500	5,320
2970	Maximum	"	820	6,600	7,420
3100	Refrigerated cases				
3120	Dairy products				
3140	Multi deck type	L.F.	33.00	820	853
3160	For rear sliding doors, add	"			86.00
3180	Delicatessen case, service deli				

		UNIT	LABOR	MAT.	TOTAL
11400.10	**FOOD SERVICE EQUIPMENT, Cont'd...**				
3190	Single deck	L.F.	250	640	890
3200	Multi deck	"	310	670	980
3220	Meat case				
3230	Single deck	L.F.	290	550	840
3240	Multi deck	"	310	600	910
3260	Produce case				
3270	Single deck	L.F.	290	640	930
3280	Multi deck	"	310	680	990
3300	Bottle coolers				
3310	6' long				
3320	Minimum	EA.	990	1,670	2,660
3340	Maximum	"	990	4,400	5,390
3345	10' long				
3350	Minimum	EA.	1,650	2,080	3,730
3360	Maximum	"	1,650	5,550	7,200
3420	Frozen food cases				
3440	Chest type	L.F.	290	750	1,040
3460	Reach-in, glass door	"	310	1,380	1,690
3470	Island case, single	"	290	710	1,000
3480	Multi deck	"	310	1,480	1,790
3500	Ice storage bins				
3520	500 lb capacity	EA.	710	2,290	3,000
3530	1000 lb capacity	"	1,410	3,200	4,610
11450.10	**RESIDENTIAL EQUIPMENT**				
0300	Compactor, 4 to 1 compaction	EA.	120	810	930
1310	Dishwasher, built-in				
1320	2 cycles	EA.	250	820	1,070
1330	4 or more cycles	"	250	1,540	1,790
1340	Disposal				
1350	Garbage disposer	EA.	160	120	280
1360	Heaters, electric, built-in				
1362	Ceiling type	EA.	160	210	370
1364	Wall type				
1370	Minimum	EA.	120	77.00	197
1374	Maximum	"	160	240	400
1400	Hood for range, 2-speed, vented				
1420	30" wide	EA.	160	130	290
1440	42" wide	"	160	770	930
1460	Ice maker, automatic				
1480	30 lb per day	EA.	71.00	1,330	1,401
1500	50 lb per day	"	250	1,760	2,010
2000	Ranges electric				
2040	Built-in, 30", 1 oven	EA.	160	1,670	1,830
2050	2 oven	"	160	2,030	2,190
2060	Counter top, 4 burner, standard	"	120	910	1,030
2070	With grill	"	120	2,500	2,620
2198	Free standing, 21", 1 oven	"	160	830	990
2200	30", 1 oven	"	99.00	1,300	1,399
2220	2 oven	"	99.00	3,300	3,399
3600	Water softener				
3620	30 grains per gallon	EA.	160	940	1,100

		UNIT	LABOR	MAT.	TOTAL
11450.10	**RESIDENTIAL EQUIPMENT, Cont'd...**				
3640	70 grains per gallon	EA.	250	1,130	1,380
11600.10	**LABORATORY EQUIPMENT**				
1000	Cabinets, base				
1020	Minimum	L.F.	40.25	170	210
1040	Maximum	"	40.25	300	340
1080	Full storage, 7' high				
1100	Minimum	L.F.	40.25	240	280
1140	Maximum	"	40.25	340	380
1150	Wall				
1160	Minimum	L.F.	48.25	100	148
1200	Maximum	"	48.25	190	238
1220	Counter tops				
1240	Minimum	S.F.	6.03	29.75	35.78
1260	Average	"	6.89	59.00	65.89
1280	Maximum	"	8.04	87.00	95.04
1300	Tables				
1320	Open underneath	S.F.	24.25	150	174
1330	Doors underneath	"	30.25	310	340
2000	Medical laboratory equipment				
2010	Analyzer				
2020	Chloride	EA.	24.75	720	745
2060	Blood	"	41.25	4,380	4,421
2070	Bath, water, utility, countertop unit	"	49.25	410	459
2080	Hot plate, lab, countertop	"	44.75	190	235
2100	Stirrer	"	44.75	290	335
2120	Incubator, anaerobic, 23x23x36"	"	250	9,890	10,140
2140	Dry heat bath	"	82.00	650	732
2160	Incinerator, for sterilizing	"	4.93	310	315
2170	Meter, serum protein	"	6.17	1,090	1,096
2180	Ph analog, general purpose	"	7.05	760	767
2190	Refrigerator, blood bank, undercounter type 153 litres	"	82.00	5,590	5,672
2200	5.4 cf, undercounter type	"	82.00	5,290	5,372
2210	Refrigerator/freezer, 4.4 cf, undercounter type	"	82.00	630	712
2220	Sealer, impulse, free standing, 20x12x4"	"	16.50	320	337
2240	Timer, electric, 1-60 minutes, bench or wall mounted	"	27.50	77.00	105
2260	Glassware washer - dryer, undercounter	"	620	6,210	6,830
2300	Balance, torsion suspension, tabletop, 4.5 lb capacity	"	27.50	580	608
2340	Binocular microscope, with in base illuminator	"	19.00	3,240	3,259
2400	Centrifuge, table model, 19x16x13"	"	19.75	1,650	1,670
2420	Clinical model, with four place head	"	11.00	720	731
11700.10	**MEDICAL EQUIPMENT**				
1000	Hospital equipment, lights				
1020	Examination, portable	EA.	41.25	1,190	1,231
1200	Meters				
1220	Air flow meter	EA.	27.50	80.00	108
1240	Oxygen flow meters	"	20.50	57.00	77.50
1900	Physical therapy				
1930	Chair, hydrotherapy	EA.	8.04	440	448
1940	Diathermy, shortwave, portable, on casters	"	19.25	3,300	3,319
1950	Exercise bicycle, floor standing, 35" x 15"	"	16.00	880	896
1960	Hydrocollator, 4 pack, portable, 129 x 90 x 160"	"	6.89	400	407

		UNIT	LABOR	MAT.	TOTAL
11700.10	**MEDICAL EQUIPMENT, Cont'd...**				
1970	Lamp, infrared, mobile with variable heat control	EA.	37.25	560	597
1980	Ultra violet, base mounted	"	37.25	500	537
2070	Stimulator, galvanic-faradic, hand held	"	3.21	100	103
2080	Ultrasound, muscle stimulator, portable, 13x13x8"	"	4.02	3,550	3,554
2120	Whirlpool, 85 gallon	"	240	3,840	4,080
2141	65 gallon capacity	"	240	3,350	3,590
2200	Radiology				
2280	Radiographic table, motor driven tilting table	EA.	4,830	30,180	35,010
2290	Fluoroscope image/tv system	"	9,660	10,180	19,840
2300	Processor for washing and drying radiographs				
2303	Water filter unit, 30" x 48-1/2" x 37-1/2"	EA.	820	35.75	856
2400	Steam sterilizers				
2410	For heat and moisture stable materials	EA.	49.25	4,280	4,329
2440	For fast drying after sterilization	"	62.00	3,300	3,362
2450	Compact unit	"	62.00	1,680	1,742
2460	Semi-automatic	"	250	6,940	7,190
2465	Floor loading				
2470	Single door	EA.	410	126,070	126,480
2480	Double door	"	490	158,960	159,450
2490	Utensil washer, sanitizer	"	380	8,630	9,010
2500	Automatic washer/sterilizer	"	990	27,410	28,400
2510	16 x 16 x 26", including generator & accessories	"	1,650	31,460	33,110
2520	Steam generator, elec., 10 kw to 180 kw	"	990	35,750	36,740
2540	Surgical scrub				
2560	Minimum	EA.	160	2,700	2,860
2580	Maximum	"	160	5,660	5,820
2600	Gas sterilizers				
2620	Automatic, free standing, 21x19x29"	EA.	490	5,660	6,150
2720	Surgical lights, ceiling mounted				
2740	Minimum	EA.	820	3,100	3,920
2760	Maximum	"	990	9,440	10,430
2800	Water stills				
2900	4 liters/hr	EA.	160	4,180	4,340
2920	8 liters/hr	"	160	4,760	4,920
2940	19 liters/hr	"	410	11,000	11,410
3060	X-ray equipment				
3070	Mobile unit				
3080	Minimum	EA.	250	6,160	6,410
3100	Maximum	"	490	31,840	32,330
3220	Minimum	"	490	4,730	5,220
3240	Maximum	"	490	9,420	9,910
3300	Incubators				
3320	15 cf	EA.	250	4,640	4,890
3330	29 cf	"	410	7,580	7,990
3340	Infant transport, portable	"	260	11,570	11,830
3460	Headwall				
3465	Aluminum, with back frame and console	EA.	250	4,040	4,290
6000	Hospital ground detection system				
6010	Power ground module	EA.	140	860	1,000
6020	Ground slave module	"	110	600	710
6030	Master ground module	"	93.00	370	463
6040	Remote indicator	"	99.00	360	459

		UNIT	LABOR	MAT.	TOTAL
11700.10	**MEDICAL EQUIPMENT, Cont'd...**				
6050	X-ray indicator	EA.	110	400	510
6060	Micro ammeter	"	120	1,220	1,340
6070	Supervisory module	"	110	1,500	1,610
6080	Ground cords	"	18.25	100	118
6100	Hospital isolation monitors, 5 ma				
6110	120v	EA.	210	2,620	2,830
6120	208v	"	210	2,510	2,720
6130	240v	"	210	2,510	2,720
6210	Digital clock-timers separate display	"	99.00	1,090	1,189
6220	One display	"	99.00	910	1,009
6230	Remote control	"	77.00	530	607
6240	Battery pack	"	77.00	120	197
6310	Surgical chronometer clock and 3 timers	"	150	1,970	2,120
6320	Auxilary control	"	72.00	710	782

TABLE OF CONTENTS   PAGE

12302.10 - CASEWORK   **12-2**
12390.10 - COUNTER TOPS   **12-2**
12500.10 - WINDOW TREATMENT   **12-2**
12510.10 - BLINDS   **12-2**

		UNIT	LABOR	MAT.	TOTAL
12302.10	**CASEWORK**				
0080	Kitchen base cabinet, prefinished, 24" deep, 35" high				
0100	12" wide	EA.	48.25	170	218
0120	18" wide	"	48.25	240	288
0140	24" wide	"	54.00	280	334
0160	27" wide	"	54.00	310	364
0180	36" wide	"	60.00	370	430
0200	48" wide	"	60.00	420	480
0220	Corner cabinet, 36" wide	"	60.00	440	500
4000	Wall cabinet, 12" deep, 12" high				
4020	30" wide	EA.	48.25	170	218
4060	36" wide	"	48.25	190	238
4110	24" high				
4120	30" wide	EA.	54.00	220	274
4140	36" wide	"	54.00	170	224
4150	30" high				
4160	12" wide	EA.	60.00	140	200
4200	24" wide	"	60.00	220	280
4320	30" wide	"	69.00	250	319
4340	36" wide	"	69.00	280	349
4350	Corner cabinet, 30" high				
4360	24" wide	EA.	80.00	230	310
4390	36" wide	"	80.00	300	380
5020	Wardrobe	"	120	960	1,080
6980	Vanity with top, laminated plastic				
7000	24" wide	EA.	120	450	570
7040	36" wide	"	160	570	730
7060	48" wide	"	190	650	840
12390.10	**COUNTER TOPS**				
1020	Stainless steel, counter top, with backsplash	S.F.	12.00	140	152
2000	Acid-proof, kemrock surface	"	8.04	46.25	54.29
12500.10	**WINDOW TREATMENT**				
1000	Drapery tracks, wall or ceiling mounted				
1040	Basic traverse rod				
1080	50 to 90"	EA.	24.25	61.00	85.25
1100	84 to 156"	"	26.75	67.00	93.75
1120	136 to 250"	"	26.75	86.00	113
1140	165 to 312"	"	30.25	140	170
1160	Traverse rod with stationary curtain rod				
1180	30 to 50"	EA.	24.25	55.00	79.25
1200	50 to 90"	"	24.25	68.00	92.25
1220	84 to 156"	"	26.75	99.00	126
1240	136 to 250"	"	30.25	110	140
1260	Double traverse rod				
1280	30 to 50"	EA.	24.25	66.00	90.25
1300	50 to 84"	"	24.25	86.00	110
1320	84 to 156"	"	26.75	110	137
1340	136 to 250"	"	30.25	130	160
12510.10	**BLINDS**				
0990	Venetian blinds				
1000	2" slats	S.F.	1.20	7.97	9.17
1020	1" slats	"	1.20	9.35	10.55

TABLE OF CONTENTS   PAGE

13056.10 - VAULTS   **13-2**

13121.10 - PRE-ENGINEERED BUILDINGS   **13-2**

13152.10 - SWIMMING POOL EQUIPMENT   **13-3**

13152.20 - SAUNAS   **13-3**

DIVISION # 13 SPECIAL

		UNIT	LABOR	MAT.	TOTAL
13056.10	**VAULTS**				
1000	Floor safes				
1010	1.0 cf	EA.	40.25	770	810
1020	1.3 cf	"	60.00	900	960
13121.10	**PRE-ENGINEERED BUILDINGS**				
1080	Pre-engineered metal building, 40'x100'				
1100	14' eave height	S.F.	4.23	13.25	17.48
1120	16' eave height	"	4.88	13.75	18.63
1140	20' eave height	"	6.34	15.00	21.34
1150	60'x100'				
1160	14' eave height	S.F.	4.23	9.84	14.07
1180	16' eave height	"	4.88	10.25	15.13
1190	20' eave height	"	6.34	11.75	18.09
1195	80'x100'				
1200	14' eave height	S.F.	4.23	10.50	14.73
1210	16' eave height	"	4.88	11.50	16.38
1220	20' eave height	"	6.34	12.75	19.09
1280	100'x100'				
1300	14' eave height	S.F.	4.23	13.25	17.48
1320	16' eave height	"	4.88	13.00	17.88
1340	20' eave height	"	6.34	15.25	21.59
1350	100'x150'				
1360	14' eave height	S.F.	4.23	11.25	15.48
1380	16' eave height	"	4.88	11.75	16.63
1400	20' eave height	"	6.34	12.50	18.84
1410	120'x150'				
1420	14' eave height	S.F.	4.23	13.25	17.48
1440	16' eave height	"	4.88	13.75	18.63
1460	20' eave height	"	6.34	15.00	21.34
1480	140'x150'				
1500	14' eave height	S.F.	4.23	11.50	15.73
1520	16' eave height	"	4.88	11.75	16.63
1540	20' eave height	"	6.34	12.50	18.84
1600	160'x200'				
1620	14' eave height	S.F.	4.23	11.00	15.23
1640	16' eave height	"	4.88	11.50	16.38
1680	20' eave height	"	6.34	11.50	17.84
1690	200'x200'				
1700	14' eave height	S.F.	4.23	10.50	14.73
1720	16' eave height	"	4.88	10.75	15.63
1740	20' eave height	"	6.34	11.75	18.09
5020	Hollow metal door and frame, 6' x 7'	EA.			1,590
5030	Sectional steel overhead door, manually operated				
5040	8' x 8'	EA.			1,720
5080	12' x 12'	"			2,340
5100	Roll-up steel door, manually operated				
5120	10' x 10'	EA.			1,300
5140	12' x 12'	"			1,510
5160	For gravity ridge ventilator with birdscreen	"			1,020
5161	9" throat x 10'	"			1,080
5181	12" throat x 10'	"			1,140
5200	For 20" rotary vent with damper	"			290

		UNIT	LABOR	MAT.	TOTAL
13121.10	**PRE-ENGINEERED BUILDINGS, Cont'd...**				
5220	For 4' x 3' fixed louver	EA.			260
5240	For 4' x 3' aluminum sliding window	"			280
5260	For 3' x 9' fiberglass panels	"			300
8020	Liner panel, 26 ga, painted steel	S.F.	1.34	2.75	4.09
8040	Wall panel insulated, 26 ga. steel, foam core	"	1.34	8.03	9.37
8060	Roof panel, 26 ga. painted steel	"	0.76	2.56	3.32
8080	Plastic (sky light)	"	0.76	5.44	6.20
9000	Insulation, 3-1/2" thick blanket, R11	"	0.35	1.76	2.11
13152.10	**SWIMMING POOL EQUIPMENT**				
1100	Diving boards				
1110	14' long				
1120	Aluminum	EA.	210	2,500	2,710
1140	Fiberglass	"	210	1,430	1,640
1500	Ladders, heavy duty				
1510	2 steps				
1520	Minimum	EA.	75.00	390	465
1540	Maximum	"	75.00	900	975
1550	4 steps				
1560	Minimum	EA.	94.00	730	824
1580	Maximum	"	94.00	1,180	1,274
1600	Lifeguard chair				
1620	Minimum	EA.	380	2,750	3,130
1640	Maximum	"	380	3,800	4,180
1700	Lights, underwater				
1720	12 volt, with transformer	EA.	94.00	260	354
1730	110 volt				
1740	Minimum	EA.	94.00	380	474
1760	Maximum	"	94.00	580	674
1780	Ground fault interrupter for 110 volt, each light	"	31.25	170	201
2000	Pool covers				
2020	Reinforced polyethylene	S.F.	2.88	0.70	3.58
2030	Vinyl water tube				
2040	Minimum	S.F.	2.88	0.88	3.76
2060	Maximum	"	2.88	1.48	4.36
2100	Slides with water tube				
2120	Minimum	EA.	310	4,050	4,360
2140	Maximum	"	310	6,820	7,130
13152.20	**SAUNAS**				
0010	Prefabricated, cedar siding, insulated panels, prehung door,				
0020	4'x8"x4'-8"x6'-6"	EA.			3,080
0030	5'-8"x6'-8"x6'-6"	"			3,960
0040	6'-8"x6'-8"x6'-6"	"			4,620
0050	7'-8"x7'-8"x6'-6"	"			7,150
0060	7'-8"x9'-8"x6'-6"	"			9,300

Design & Construction Resources

TABLE OF CONTENTS   PAGE

14210.10 - ELEVATORS   14-2

14300.10 - ESCALATORS   14-2

14410.10 - PERSONNEL LIFTS   14-2

14450.10 - VEHICLE LIFTS   14-3

14560.10 - CHUTES   14-3

14580.10 - PNEUMATIC SYSTEMS   14-3

14580.30 - DUMBWAITERS   14-3

		UNIT	LABOR	MAT.	TOTAL

14210.10 ELEVATORS

		UNIT	LABOR	MAT.	TOTAL
0120	Passenger elevators, electric, geared				
0502	Based on a shaft of 6 stops and 6 openings				
0510	50 fpm, 2000 lb	EA.	2,110	105,450	107,560
0520	100 fpm, 2000 lb	"	2,350	109,350	111,700
0525	150 fpm				
0530	2000 lb	EA.	2,640	121,070	123,710
0550	3000 lb	"	3,020	152,310	155,330
0560	4000 lb	"	3,520	158,170	161,690
1002	Based on a shaft of 8 stops and 8 openings				
1010	300 fpm				
1020	3000 lb	EA.	4,230	181,610	185,840
1040	3500 lb	"	4,230	185,670	189,900
1060	4000 lb	"	4,700	195,280	199,980
1080	5000 lb	"	5,030	218,710	223,740
1502	Hydraulic, based on a shaft of 3 stops, 3 openings				
1508	50 fpm				
1510	2000 lb	EA.	1,760	69,010	70,770
1520	2500 lb	"	1,760	73,860	75,620
1530	3000 lb	"	1,840	77,930	79,770
1600	For each additional; 50 fpm add per stop, $3500				
1620	500 lb, add per stop, $3500				
1630	Opening, add, $4200				
1640	Stop, add per stop, $5300				
1660	Bonderized steel door, add per opening, $400				
1670	Colored aluminum door, add per opening, $1500				
1680	Stainless steel door, add per opening, $650				
1690	Cast bronze door, add per opening, $1200				
1730	Custom cab interior, add per cab, $5000				
2000	Small elevators, 4 to 6 passenger capacity				
2005	Electric, push				
2010	2 stops	EA.	1,760	25,200	26,960
2020	3 stops	"	1,920	31,010	32,930
2030	4 stops	"	2,110	35,860	37,970

14300.10 ESCALATORS

		UNIT	LABOR	MAT.	TOTAL
1000	Escalators				
1020	32" wide, floor to floor				
1040	12' high	EA.	3,520	137,320	140,840
1050	15' high	"	4,230	148,930	153,160
1060	18' high	"	5,280	160,540	165,820
1070	22' high	"	7,050	174,070	181,120
1080	25' high	"	8,450	193,420	201,870
1085	48" wide				
1090	12' high	EA.	3,640	152,800	156,440
1100	15' high	"	4,400	166,340	170,740
1120	18' high	"	5,560	177,940	183,500
1130	22' high	"	7,550	199,220	206,770
1140	25' high	"	8,450	212,660	221,110

14410.10 PERSONNEL LIFTS

		UNIT	LABOR	MAT.	TOTAL
1000	Electrically operated, 1 or 2 person lift				
1001	With attached foot platforms				
1020	3 stops	EA.			24,840

		UNIT	LABOR	MAT.	TOTAL
14410.10	**PERSONNEL LIFTS, Cont'd...**				
1040	5 stops	EA.			30,530
1060	7 stops	"			35,380
2000	For each additional stop, add $1250				
3020	Residential stair climber, per story	EA.	410	6,250	6,660
14450.10	**VEHICLE LIFTS**				
1020	Automotive hoist, one post, semi-hydraulic, 8,000 lb	EA.	2,110	4,900	7,010
1040	Full hydraulic, 8,000 lb	"	2,110	4,620	6,730
1060	2 post, semi-hydraulic, 10,000 lb	"	3,020	4,950	7,970
1070	Full hydraulic				
1080	10,000 lb	EA.	3,020	3,410	6,430
1100	13,000 lb	"	5,280	6,640	11,920
1120	18,500 lb	"	5,280	8,460	13,740
1140	24,000 lb	"	5,280	12,780	18,060
1160	26,000 lb	"	5,280	14,300	19,580
1170	Pneumatic hoist, fully hydraulic				
1180	11,000 lb	EA.	7,050	8,990	16,040
1200	24,000 lb	"	7,050	13,970	21,020
14560.10	**CHUTES**				
1020	Linen chutes, stainless steel, with supports				
1030	18" dia.	L.F.	3.83	130	134
1040	24" dia.	"	4.13	170	174
1050	30" dia.	"	4.47	180	184
1060	Hopper	EA.	35.75	2,100	2,136
1070	Skylight	"	54.00	1,270	1,324
14580.10	**PNEUMATIC SYSTEMS**				
6000	Air-lift conveyor, 115 Volt, single phase, 100' long, 3" carrier	EA.	4,490	3,190	7,680
6010	4" carrier	"	4,490	4,070	8,560
6020	6" carrier	"	4,490	7,590	12,080
7000	Pneumatic tube system accessories				
7010	Couplings and hanging accessories for 3" carrier system	L.F.			4.67
7020	For 4" carrier system	"			5.83
7030	For 6" carrier system	"			5.83
7040	24" CL. Expanded 90° bend, heavy duty, 3"-6" carrier systems	EA.			110
7050	36" CL. Expanded 90° bend, heavy duty, 3"-6" carrier systems	"			240
7060	48" CL. Expanded 45° bend, heavy duty, 3" carrier system	"			83.00
7070	4" carrier system	"			110
7080	6" carrier system	"			160
7090	48" CL. Expanded 90° bend, heavy duty, 3" carrier system	"			100
8000	4" carrier system	"			140
8010	6" carrier system	"			240
8020	Stainless steel body up-grade, 3" carrier system	"			340
8030	4" carrier system	"			350
8040	6" carrier system	"			470
14580.30	**DUMBWAITERS**				
0010	28' travel, extruded alum., 4 stops, 100 lbs. capacity	EA.			4,620
0020	150 lbs. capacity	"			5,890
0030	200 lbs. capacity	"			9,230

Design & Construction Resources

TABLE OF CONTENTS   PAGE

15120.10 - BACKFLOW PREVENTERS	**15-2**
15410.06 - C.I. PIPE, BELOW GROUND	15-2
15410.10 - COPPER PIPE	**15-2**
15410.11 - COPPER FITTINGS	15-2
15410.82 - GALVANIZED STEEL PIPE	**15-3**
15430.23 - CLEANOUTS	15-3
15430.25 - HOSE BIBBS	**15-3**
15430.60 - VALVES	15-3
15430.65 - VACUUM BREAKERS	**15-4**
15430.68 - STRAINERS	15-4
15430.70 - DRAINS, ROOF & FLOOR	**15-4**
15440.10 - BATHS	15-4
15440.12 - DISPOSALS & ACCESSORIES	**15-5**
15440.15 - FAUCETS	15-5
15440.18 - HYDRANTS	**15-6**
15440.20 - LAVATORIES	15-6
15440.30 - SHOWERS	**15-6**
15440.40 - SINKS	15-6
15440.60 - WATER CLOSETS	**15-7**
15440.70 - WATER HEATERS	15-7
15780.20 - ROOFTOP UNITS	**15-8**
15830.70 - UNIT HEATERS	15-8
15855.10 - AIR HANDLING UNITS	**15-8**
15870.20 - EXHAUST FANS	15-8
15890.10 - METAL DUCTWORK	**15-8**
15890.30 - FLEXIBLE DUCTWORK	15-9
15910.10 - DAMPERS	**15-9**
15940.10 - DIFFUSERS	**15-9**

		UNIT	LABOR	MAT.	TOTAL
15120.10	**BACKFLOW PREVENTERS**				
0080	Backflow preventer, flanged, cast iron, with valves				
0100	3" pipe	EA.	270	2,760	3,030
15410.06	**C.I. PIPE, BELOW GROUND**				
1010	No hub pipe				
1020	1-1/2" pipe	L.F.	2.66	6.18	8.84
1030	2" pipe	"	2.96	6.34	9.30
1120	3" pipe	"	3.33	8.75	12.08
1220	4" pipe	"	4.44	11.25	15.69
15410.10	**COPPER PIPE**				
0880	Type "K" copper				
0900	1/2"	L.F.	1.66	5.50	7.16
1000	3/4"	"	1.77	10.00	11.77
1020	1"	"	1.90	12.75	14.65
3000	DWV, copper				
3020	1-1/4"	L.F.	2.22	10.25	12.47
3030	1-1/2"	"	2.42	13.00	15.42
3040	2"	"	2.66	14.75	17.41
3070	3"	"	2.96	27.75	30.71
3080	4"	"	3.33	45.75	49.08
3090	6"	"	3.80	180	184
6080	Type "L" copper				
6090	1/4"	L.F.	1.56	3.35	4.91
6095	3/8"	"	1.56	3.63	5.19
6100	1/2"	"	1.66	5.50	7.16
6190	3/4"	"	1.77	6.27	8.04
6240	1"	"	1.90	8.47	10.37
6580	Type "M" copper				
6600	1/2"	L.F.	1.66	3.57	5.23
6620	3/4"	"	1.77	5.00	6.77
6630	1"	"	1.90	6.49	8.39
15410.11	**COPPER FITTINGS**				
0850	Slip coupling				
0860	1/4"	EA.	17.75	0.79	18.54
0870	1/2"	"	21.25	1.33	22.58
0880	3/4"	"	26.75	2.77	29.52
0890	1"	"	29.50	5.84	35.34
2660	Street ells, copper				
2670	1/4"	EA.	21.25	7.35	28.60
2680	3/8"	"	24.25	5.08	29.33
2690	1/2"	"	26.75	2.04	28.79
2700	3/4"	"	28.00	4.31	32.31
2710	1"	"	29.50	11.25	40.75
4190	DWV fittings, coupling with stop				
4210	1-1/2"	EA.	33.25	7.50	40.75
4230	2"	"	35.50	10.50	46.00
4260	3"	"	44.50	20.25	64.75
4280	3" x 2"	"	44.50	46.00	90.50
4290	4"	"	53.00	64.00	117
4300	Slip coupling				
4310	1-1/2"	EA.	33.25	11.75	45.00
4320	2"	"	35.50	12.75	48.25

		UNIT	LABOR	MAT.	TOTAL
15410.11	**COPPER FITTINGS, Cont'd...**				
4330	3"	EA.	44.50	18.00	62.50
4340	90 ells				
4350	1-1/2"	EA.	33.25	19.25	52.50
4360	1-1/2" x 1-1/4"	"	33.25	38.00	71.25
4370	2"	"	35.50	25.25	60.75
4380	2" x 1-1/2"	"	35.50	30.50	66.00
4390	3"	"	44.50	62.00	107
4400	4"	"	53.00	300	353
4410	Street, 90 elbows				
4420	1-1/2"	EA.	33.25	18.00	51.25
4430	2"	"	35.50	34.75	70.25
4440	3"	"	44.50	74.00	119
4450	4"	"	53.00	250	303
5410	No-hub adapters				
5420	1-1/2" x 2"	EA.	33.25	32.50	65.75
5430	2"	"	35.50	30.50	66.00
5440	2" x 3"	"	35.50	70.00	106
5450	3"	"	44.50	62.00	107
5460	3" x 4"	"	44.50	130	175
5470	4"	"	53.00	130	183
15410.82	**GALVANIZED STEEL PIPE**				
1000	Galvanized pipe				
1020	1/2" pipe	L.F.	5.33	2.53	7.86
1040	3/4" pipe	"	6.66	3.30	9.96
1200	90 degree ell, 150 lb malleable iron, galvanized				
1210	1/2"	EA.	10.75	2.69	13.44
1220	3/4"	"	13.25	3.57	16.82
1400	45 degree ell, 150 lb m.i., galv.				
1410	1/2"	EA.	10.75	4.29	15.04
1420	3/4"	"	13.25	5.83	19.08
1520	Tees, straight, 150 lb m.i., galv.				
1530	1/2"	EA.	13.25	3.57	16.82
1540	3/4"	"	15.25	5.94	21.19
1800	Couplings, straight, 150 lb m.i., galv.				
1810	1/2"	EA.	10.75	3.30	14.05
1820	3/4"	"	11.75	3.96	15.71
15430.23	**CLEANOUTS**				
0980	Cleanout, wall				
1000	2"	EA.	35.50	160	196
1020	3"	"	35.50	230	266
1040	4"	"	44.50	240	285
1050	Floor				
1060	2"	EA.	44.50	150	195
1080	3"	"	44.50	190	235
1100	4"	"	53.00	200	253
15430.25	**HOSE BIBBS**				
0005	Hose bibb				
0010	1/2"	EA.	17.75	9.07	26.82
0200	3/4"	"	17.75	9.62	27.37

		UNIT	LABOR	MAT.	TOTAL
15430.60	**VALVES**				
0780	Gate valve, 125 lb, bronze, soldered				
0800	1/2"	EA.	13.25	25.75	39.00
1000	3/4"	"	13.25	51.00	64.25
1280	Check valve, bronze, soldered, 125 lb				
1300	1/2"	EA.	13.25	36.75	50.00
1320	3/4"	"	13.25	45.75	59.00
1790	Globe valve, bronze, soldered, 125 lb				
1800	1/2"	EA.	15.25	55.00	70.25
1810	3/4"	"	16.75	66.00	82.75
15430.65	**VACUUM BREAKERS**				
1000	Vacuum breaker, atmospheric, threaded connection				
1010	3/4"	EA.	21.25	45.75	67.00
1018	Anti-siphon, brass				
1020	3/4"	EA.	21.25	52.00	73.25
15430.68	**STRAINERS**				
0980	Strainer, Y pattern, 125 psi, cast iron body, threaded				
1000	3/4"	EA.	19.00	10.75	29.75
1980	250 psi, brass body, threaded				
2000	3/4"	EA.	21.25	29.75	51.00
2130	Cast iron body, threaded				
2140	3/4"	EA.	21.25	17.50	38.75
15430.70	**DRAINS, ROOF & FLOOR**				
1020	Floor drain, cast iron, with cast iron top				
1030	2"	EA.	44.50	120	165
1040	3"	"	44.50	140	185
1050	4"	"	44.50	290	335
1090	Roof drain, cast iron				
1100	2"	EA.	44.50	210	255
1110	3"	"	44.50	260	305
1120	4"	"	44.50	270	315
15440.10	**BATHS**				
0980	Bath tub, 5' long				
1000	Minimum	EA.	180	530	710
1020	Average	"	270	1,160	1,430
1040	Maximum	"	530	2,640	3,170
1050	6' long				
1060	Minimum	EA.	180	590	770
1080	Average	"	270	1,210	1,480
1100	Maximum	"	530	3,420	3,950
1110	Square tub, whirlpool, 4'x4'				
1120	Minimum	EA.	270	1,810	2,080
1140	Average	"	530	2,570	3,100
1160	Maximum	"	670	7,850	8,520
1170	5'x5'				
1180	Minimum	EA.	270	1,810	2,080
1200	Average	"	530	2,570	3,100
1220	Maximum	"	670	8,000	8,670
1230	6'x6'				
1240	Minimum	EA.	270	2,210	2,480
1260	Average	"	530	3,230	3,760
1280	Maximum	"	670	9,270	9,940

DIVISION # 15 MECHANICAL

		UNIT	LABOR	MAT.	TOTAL
15440.10	**BATHS, Cont'd...**				
8980	For trim and rough-in				
9000	Minimum	EA.	180	190	370
9020	Average	"	270	280	550
9040	Maximum	"	530	780	1,310
15440.12	**DISPOSALS & ACCESSORIES**				
0040	Continuous feed				
0050	Minimum	EA.	110	72.00	182
0060	Average	"	130	200	330
0070	Maximum	"	180	390	570
0200	Batch feed, 1/2 hp				
0220	Minimum	EA.	110	280	390
0230	Average	"	130	550	680
0240	Maximum	"	180	950	1,130
1100	Hot water dispenser				
1110	Minimum	EA.	110	200	310
1120	Average	"	130	320	450
1130	Maximum	"	180	510	690
1140	Epoxy finish faucet	"	110	290	400
1160	Lock stop assembly	"	67.00	61.00	128
1170	Mounting gasket	"	44.50	7.04	51.54
1180	Tailpipe gasket	"	44.50	1.03	45.53
1190	Stopper assembly	"	53.00	24.00	77.00
1200	Switch assembly, on/off	"	89.00	27.50	117
1210	Tailpipe gasket washer	"	26.75	1.10	27.85
1220	Stop gasket	"	29.50	2.42	31.92
1230	Tailpipe flange	"	26.75	0.27	27.02
1240	Tailpipe	"	33.25	3.13	36.38
15440.15	**FAUCETS**				
0980	Kitchen				
1000	Minimum	EA.	89.00	130	219
1020	Average	"	110	230	340
1040	Maximum	"	130	290	420
1050	Bath				
1060	Minimum	EA.	89.00	130	219
1080	Average	"	110	240	350
1100	Maximum	"	130	370	500
1110	Lavatory, domestic				
1120	Minimum	EA.	89.00	140	229
1140	Average	"	110	280	390
1160	Maximum	"	130	460	590
1290	Washroom				
1300	Minimum	EA.	89.00	160	249
1320	Average	"	110	280	390
1340	Maximum	"	130	510	640
1350	Handicapped				
1360	Minimum	EA.	110	200	310
1380	Average	"	130	360	490
1400	Maximum	"	180	560	740
1410	Shower				
1420	Minimum	EA.	89.00	160	249
1440	Average	"	110	320	430

DIVISION # 15 MECHANICAL

		UNIT	LABOR	MAT.	TOTAL
15440.15	**FAUCETS, Cont'd...**				
1460	Maximum	EA.	130	510	640
1480	For trim and rough-in				
1500	Minimum	EA.	110	79.00	189
1520	Average	"	130	120	250
1540	Maximum	"	270	200	470
15440.18	**HYDRANTS**				
0980	Wall hydrant				
1000	8" thick	EA.	89.00	340	429
1020	12" thick	"	110	400	510
15440.20	**LAVATORIES**				
1980	Lavatory, counter top, porcelain enamel on cast iron				
2000	Minimum	EA.	110	190	300
2010	Average	"	130	290	420
2020	Maximum	"	180	520	700
2080	Wall hung, china				
2100	Minimum	EA.	110	260	370
2110	Average	"	130	310	440
2120	Maximum	"	180	770	950
2280	Handicapped				
2300	Minimum	EA.	130	430	560
2310	Average	"	180	500	680
2320	Maximum	"	270	830	1,100
8980	For trim and rough-in				
9000	Minimum	EA.	130	220	350
9020	Average	"	180	370	550
9040	Maximum	"	270	460	730
15440.30	**SHOWERS**				
0980	Shower, fiberglass, 36"x34"x84"				
1000	Minimum	EA.	380	570	950
1020	Average	"	530	800	1,330
1040	Maximum	"	530	1,160	1,690
2980	Steel, 1 piece, 36"x36"				
3000	Minimum	EA.	380	530	910
3020	Average	"	530	800	1,330
3040	Maximum	"	530	950	1,480
3980	Receptor, molded stone, 36"x36"				
4000	Minimum	EA.	180	220	400
4020	Average	"	270	370	640
4040	Maximum	"	440	570	1,010
8980	For trim and rough-in				
9000	Minimum	EA.	240	220	460
9020	Average	"	300	370	670
9040	Maximum	"	530	460	990
15440.40	**SINKS**				
0980	Service sink, 24"x29"				
1000	Minimum	EA.	130	640	770
1020	Average	"	180	790	970
1040	Maximum	"	270	1,170	1,440
2000	Kitchen sink, single, stainless steel, single bowl				
2020	Minimum	EA.	110	280	390
2040	Average	"	130	320	450

		UNIT	LABOR	MAT.	TOTAL
15440.40	**SINKS, Cont'd...**				
2060	Maximum	EA.	180	580	760
2070	Double bowl				
2080	Minimum	EA.	130	320	450
2100	Average	"	180	360	540
2120	Maximum	"	270	620	890
2190	Porcelain enamel, cast iron, single bowl				
2200	Minimum	EA.	110	200	310
2220	Average	"	130	260	390
2240	Maximum	"	180	410	590
2250	Double bowl				
2260	Minimum	EA.	130	280	410
2280	Average	"	180	390	570
2300	Maximum	"	270	550	820
2980	Mop sink, 24"x36"x10"				
3000	Minimum	EA.	110	480	590
3020	Average	"	130	580	710
3040	Maximum	"	180	780	960
5980	Washing machine box				
6000	Minimum	EA.	130	180	310
6040	Average	"	180	250	430
6060	Maximum	"	270	310	580
8980	For trim and rough-in				
9000	Minimum	EA.	180	290	470
9020	Average	"	270	440	710
9040	Maximum	"	360	560	920
15440.60	**WATER CLOSETS**				
0980	Water closet flush tank, floor mounted				
1000	Minimum	EA.	130	330	460
1010	Average	"	180	650	830
1020	Maximum	"	270	1,020	1,290
1030	Handicapped				
1040	Minimum	EA.	180	370	550
1050	Average	"	270	670	940
1060	Maximum	"	530	1,280	1,810
8980	For trim and rough-in				
9000	Minimum	EA.	130	210	340
9020	Average	"	180	250	430
9040	Maximum	"	270	330	600
15440.70	**WATER HEATERS**				
0980	Water heater, electric				
1000	6 gal	EA.	89.00	330	419
1020	10 gal	"	89.00	350	439
1030	15 gal	"	89.00	360	449
1040	20 gal	"	110	410	520
1050	30 gal	"	110	440	550
1060	40 gal	"	110	560	670
1070	52 gal	"	130	630	760
2980	Oil fired				
3000	20 gal	EA.	270	1,300	1,570
3020	50 gal	"	380	2,020	2,400

		UNIT	LABOR	MAT.	TOTAL
15610.10	**FURNACES**				
0980	Electric, hot air				
1000	40 mbh	EA.	270	810	1,080
1020	60 mbh	"	280	880	1,160
1040	80 mbh	"	300	960	1,260
1060	100 mbh	"	310	1,080	1,390
1080	125 mbh	"	320	1,320	1,640
1980	Gas fired hot air				
2000	40 mbh	EA.	270	810	1,080
2020	60 mbh	"	280	870	1,150
2040	80 mbh	"	300	1,000	1,300
2060	100 mbh	"	310	1,040	1,350
2080	125 mbh	"	320	1,140	1,460
2980	Oil fired hot air				
3000	40 mbh	EA.	270	1,090	1,360
3020	60 mbh	"	280	1,480	1,760
3040	80 mbh	"	300	1,590	1,890
3060	100 mbh	"	310	1,810	2,120
3080	125 mbh	"	320	1,870	2,190
15780.20	**ROOFTOP UNITS**				
0980	Packaged, single zone rooftop unit, with roof curb				
1000	2 ton	EA.	530	3,400	3,930
1020	3 ton	"	530	3,580	4,110
1040	4 ton	"	670	3,900	4,570
15830.70	**UNIT HEATERS**				
0980	Steam unit heater, horizontal				
1000	12,500 btuh, 200 cfm	EA.	89.00	510	599
1010	17,000 btuh, 300 cfm	"	89.00	570	659
15855.10	**AIR HANDLING UNITS**				
0980	Air handling unit, medium pressure, single zone				
1000	1500 cfm	EA.	330	4,000	4,330
1060	3000 cfm	"	590	5,260	5,850
8980	Rooftop air handling units				
9000	4950 cfm	EA.	590	11,500	12,090
9060	7370 cfm	"	760	14,580	15,340
15870.20	**EXHAUST FANS**				
0160	Belt drive roof exhaust fans				
1020	640 cfm, 2618 fpm	EA.	67.00	1,030	1,097
1030	940 cfm, 2604 fpm	"	67.00	1,340	1,407
15890.10	**METAL DUCTWORK**				
0090	Rectangular duct				
0100	Galvanized steel				
1000	Minimum	Lb.	4.84	0.88	5.72
1010	Average	"	5.92	1.10	7.02
1020	Maximum	"	8.88	1.68	10.56
1080	Aluminum				
1100	Minimum	Lb.	10.75	2.29	13.04
1120	Average	"	13.25	3.05	16.30
1140	Maximum	"	17.75	3.79	21.54
1160	Fittings				
1180	Minimum	EA.	17.75	7.26	25.01
1200	Average	"	26.75	11.00	37.75

		UNIT	LABOR	MAT.	TOTAL
15890.10	**METAL DUCTWORK, Cont'd...**				
1220	Maximum	EA.	53.00	16.00	69.00
15890.30	**FLEXIBLE DUCTWORK**				
1010	Flexible duct, 1.25" fiberglass				
1020	5" dia.	L.F.	2.66	2.58	5.24
1040	6" dia.	"	2.96	3.13	6.09
1060	7" dia.	"	3.13	3.46	6.59
1080	8" dia.	"	3.33	3.96	7.29
1100	10" dia.	"	3.80	4.45	8.25
1120	12" dia.	"	4.10	5.50	9.60
9000	Flexible duct connector, 3" wide fabric	"	8.88	2.31	11.19
15910.10	**DAMPERS**				
0980	Horizontal parallel aluminum backdraft damper				
1000	12" x 12"	EA.	13.25	69.00	82.25
1010	16" x 16"	"	15.25	71.00	86.25
15940.10	**DIFFUSERS**				
1980	Ceiling diffusers, round, baked enamel finish				
2000	6" dia.	EA.	17.75	54.00	71.75
2020	8" dia.	"	22.25	65.00	87.25
2040	10" dia.	"	22.25	72.00	94.25
2060	12" dia.	"	22.25	92.00	114
2480	Rectangular				
2500	6x6"	EA.	17.75	58.00	75.75
2520	9x9"	"	26.75	70.00	96.75
2540	12x12"	"	26.75	100	127
2560	15x15"	"	26.75	130	157
2580	18x18"	"	26.75	160	187

Design & Construction Resources

TABLE OF CONTENTS   PAGE

16050.30 - BUS DUCT	**16-2**
16110.22 - EMT CONDUIT	**16-2**
16110.24 - GALVANIZED CONDUIT	**16-2**
16110.28 - STEEL CONDUIT	**16-3**
16110.35 - SURFACE MOUNTED RACEWAY	**16-3**
16120.43 - COPPER CONDUCTORS	**16-4**
16120.47 - SHEATHED CABLE	**16-4**
16130.60 - PULL AND JUNCTION BOXES	16-5
16130.80 - RECEPTACLES	**16-5**
16350.10 - CIRCUIT BREAKERS	16-5
16395.10 - GROUNDING	**16-5**
16430.20 - METERING	16-5
16490.10 - SWITCHES	**16-6**
16510.05 - INTERIOR LIGHTING	16-6
16510.10 - LIGHTING INDUSTRIAL	**16-6**
16710.10 - COMMUNICATIONS COMPONENTS	16-7
16760.10 - AUDIO/VIDEO CABLES	**16-8**
16850.10 - ELECTRIC HEATING	16-8
16910.40 - CONTROL CABLE	**16-8**

		UNIT	LABOR	MAT.	TOTAL
16050.30	**BUS DUCT**				
1000	Bus duct, 100a, plug-in				
1010	10', 600v	EA.	170	230	400
1020	With ground	"	260	310	570
1145	Circuit breakers, with enclosure				
1147	1 pole				
1150	15a-60a	EA.	62.00	230	292
1160	70a-100a	"	77.00	260	337
1165	2 pole				
1170	15a-60a	EA.	68.00	340	408
1180	70a-100a	"	80.00	410	490
16110.22	**EMT CONDUIT**				
0080	EMT conduit				
0100	1/2"	L.F.	1.86	0.36	2.22
1020	3/4"	"	2.46	0.68	3.14
1030	1"	"	3.08	1.07	4.15
2980	90 deg. elbow				
3000	1/2"	EA.	5.48	5.22	10.70
3040	3/4"	"	6.17	5.72	11.89
3060	1"	"	6.58	8.69	15.27
3980	Connector, steel compression				
4000	1/2"	EA.	5.48	1.17	6.65
4040	3/4"	"	5.48	2.24	7.72
4060	1"	"	5.48	3.38	8.86
0080	Flexible conduit, steel				
0100	3/8"	L.F.	1.86	0.39	2.25
1020	1/2	"	1.86	0.45	2.31
1040	3/4"	"	2.46	0.60	3.06
1060	1"	"	2.46	1.18	3.64
16110.24	**GALVANIZED CONDUIT**				
1980	Galvanized rigid steel conduit				
2000	1/2"	L.F.	2.46	1.30	3.76
2040	3/4"	"	3.08	1.51	4.59
2060	1"	"	3.65	2.35	6.00
2080	1-1/4"	"	4.93	3.35	8.28
2100	1-1/2"	"	5.48	3.92	9.40
2120	2"	"	6.17	4.65	10.82
2480	90 degree ell				
2500	1/2"	EA.	15.50	10.25	25.75
2540	3/4"	"	19.00	10.75	29.75
2560	1"	"	23.50	16.50	40.00
2580	1-1/4"	"	27.50	22.75	50.25
2590	1-1/2"	"	30.75	28.00	58.75
2600	2"	"	33.00	40.75	73.75
3200	Couplings, with set screws				
3220	1/2"	EA.	3.08	5.17	8.25
3260	3/4"	"	3.65	6.82	10.47
3280	1"	"	4.93	11.00	15.93
3300	1-1/4"	"	6.17	18.50	24.67
3320	1-1/2"	"	7.59	24.00	31.59
3340	2"	"	8.97	54.00	62.97

		UNIT	LABOR	MAT.	TOTAL
16110.25	**PLASTIC CONDUIT**				
3010	PVC conduit, schedule 40				
3020	1/2"	L.F.	1.86	0.29	2.15
3040	3/4"	"	1.86	0.33	2.19
3060	1"	"	2.46	0.49	2.95
3080	1-1/4"	"	2.46	0.69	3.15
3100	1-1/2"	"	3.08	0.80	3.88
3120	2"	"	3.08	0.96	4.04
3480	Couplings				
3500	1/2"	EA.	3.08	0.47	3.55
3520	3/4"	"	3.08	0.49	3.57
3540	1"	"	3.08	0.79	3.87
3560	1-1/4"	"	3.65	1.00	4.65
3580	1-1/2"	"	3.65	1.32	4.97
3600	2"	"	3.65	1.87	5.52
3705	90 degree elbows				
3710	1/2"	EA.	6.17	1.87	8.04
3740	3/4"	"	7.59	1.81	9.40
3760	1"	"	7.59	2.82	10.41
3780	1-1/4"	"	8.97	4.07	13.04
3800	1-1/2"	"	11.75	5.39	17.14
3810	2"	"	13.75	7.40	21.15
16110.28	**STEEL CONDUIT**				
7980	Intermediate metal conduit (IMC)				
8000	1/2"	L.F.	1.86	1.43	3.29
8040	3/4"	"	2.46	1.70	4.16
8060	1"	"	3.08	2.57	5.65
8080	1-1/4"	"	3.65	3.28	6.93
8100	1-1/2"	"	4.93	4.12	9.05
8120	2"	"	5.48	5.35	10.83
8490	90 degree ell				
8500	1/2"	EA.	15.50	9.18	24.68
8540	3/4"	"	19.00	11.25	30.25
8560	1"	"	23.50	16.25	39.75
8580	1-1/4"	"	27.50	26.25	53.75
8600	1-1/2"	"	30.75	29.50	60.25
8620	2"	"	35.25	42.25	77.50
9260	Couplings				
9280	1/2"	EA.	3.08	1.85	4.93
9290	3/4"	"	3.65	2.27	5.92
9300	1"	"	4.93	3.37	8.30
9310	1-1/4"	"	5.48	4.21	9.69
9320	1-1/2"	"	6.17	5.32	11.49
9330	2"	"	6.58	7.02	13.60
16110.35	**SURFACE MOUNTED RACEWAY**				
0980	Single Raceway				
1000	3/4" x 17/32" Conduit	L.F.	2.46	1.67	4.13
1020	Mounting Strap	EA.	3.29	0.45	3.74
1040	Connector	"	3.29	0.60	3.89
1060	Elbow				
2000	45 degree	EA.	3.08	7.62	10.70
2020	90 degree	"	3.08	2.43	5.51

DIVISION # 16 ELECTRICAL

		UNIT	LABOR	MAT.	TOTAL
16110.35	**SURFACE MOUNTED RACEWAY, Cont'd...**				
2040	internal	EA.	3.08	3.05	6.13
2050	external	"	3.08	2.82	5.90
2060	Switch	"	24.75	19.75	44.50
2100	Utility Box	"	24.75	13.25	38.00
2110	Receptacle	"	24.75	23.50	48.25
2140	3/4" x 21/32" Conduit	L.F.	2.46	1.90	4.36
2160	Mounting Strap	EA.	3.29	0.70	3.99
2180	Connector	"	3.29	0.72	4.01
2200	Elbow				
2210	45 degree	EA.	3.08	9.41	12.49
2220	90 degree	"	3.08	2.59	5.67
2240	internal	"	3.08	3.52	6.60
2260	external	"	3.08	3.52	6.60
3000	Switch	"	24.75	19.75	44.50
3010	Utility Box	"	24.75	13.25	38.00
3020	Receptacle	"	24.75	23.50	48.25
16120.43	**COPPER CONDUCTORS**				
0980	Copper conductors, type THW, solid				
1000	#14	L.F.	0.24	0.12	0.36
1040	#12	"	0.30	0.18	0.48
1060	#10	"	0.37	0.28	0.65
2010	THHN-THWN, solid				
2020	#14	L.F.	0.24	0.12	0.36
2040	#12	"	0.30	0.18	0.48
2060	#10	"	0.37	0.28	0.65
6215	Type "BX" solid armored cable				
6220	#14/2	L.F.	1.54	0.66	2.20
6230	#14/3	"	1.73	0.95	2.68
6240	#14/4	"	1.89	1.41	3.30
6250	#12/2	"	1.73	0.67	2.40
6260	#12/3	"	1.89	1.07	2.96
6270	#12/4	"	2.14	1.47	3.61
6280	#10/2	"	1.89	1.25	3.14
6290	#10/3	"	2.14	1.80	3.94
6300	#10/4	"	2.46	2.80	5.26
16120.47	**SHEATHED CABLE**				
6700	Non-metallic sheathed cable				
6705	Type NM cable with ground				
6710	#14/2	L.F.	0.92	0.35	1.27
6720	#12/2	"	0.98	0.53	1.51
6730	#10/2	"	1.09	0.85	1.94
6740	#8/2	"	1.23	1.39	2.62
6750	#6/2	"	1.54	2.20	3.74
6760	#14/3	"	1.59	0.49	2.08
6770	#12/3	"	1.64	0.77	2.41
6780	#10/3	"	1.67	1.22	2.89
6790	#8/3	"	1.70	2.05	3.75
6800	#6/3	"	1.73	3.32	5.05
6810	#4/3	"	1.97	5.84	7.81
6820	#2/3	"	2.14	8.75	10.89

		UNIT	LABOR	MAT.	TOTAL
16130.40	**BOXES**				
5000	Round cast box, type SEH				
5010	1/2"	EA.	21.50	20.00	41.50
5020	3/4"	"	26.00	20.00	46.00
16130.60	**PULL AND JUNCTION BOXES**				
1050	4"				
1060	Octagon box	EA.	7.05	2.79	9.84
1070	Box extension	"	3.65	4.64	8.29
1080	Plaster ring	"	3.65	3.08	6.73
1100	Cover blank	"	3.65	1.17	4.82
1120	Square box	"	7.05	3.38	10.43
1140	Box extension	"	3.65	4.69	8.34
1160	Plaster ring	"	3.65	2.57	6.22
1180	Cover blank	"	3.65	1.32	4.97
16130.80	**RECEPTACLES**				
0500	Contractor grade duplex receptacles, 15a 120v				
0510	Duplex	EA.	12.25	1.04	13.29
1000	125 volt, 20a, duplex, grounding type, standard grade	"	12.25	9.07	21.32
1040	Ground fault interrupter type	"	18.25	12.25	30.50
1520	250 volt, 20a, 2 pole, single receptacle, ground type	"	12.25	6.60	18.85
1540	120/208v, 4 pole, single receptacle, twist lock				
1560	20a	EA.	21.50	21.50	43.00
1580	50a	"	21.50	41.00	62.50
1590	125/250v, 3 pole, flush receptacle				
1600	30a	EA.	18.25	6.82	25.07
1620	50a	"	18.25	7.89	26.14
1640	60a	"	21.50	32.00	53.50
16350.10	**CIRCUIT BREAKERS**				
5000	Load center circuit breakers, 240v				
5010	1 pole, 10-60a	EA.	15.50	16.50	32.00
5015	2 pole				
5020	10-60a	EA.	24.75	35.25	60.00
5030	70-100a	"	41.25	100	141
5040	110-150a	"	44.75	200	245
5065	Load center, G.F.I. breakers, 240v				
5070	1 pole, 15-30a	EA.	18.25	110	128
5095	Tandem breakers, 240v				
5100	1 pole, 15-30a	EA.	24.75	31.25	56.00
5110	2 pole, 15-30a	"	33.00	57.00	90.00
16395.10	**GROUNDING**				
0500	Ground rods, copper clad, 1/2" x				
0510	6'	EA.	41.25	10.75	52.00
0520	8'	"	44.75	14.75	59.50
0535	5/8" x				
0550	6'	EA.	44.75	14.25	59.00
0560	8'	"	62.00	18.50	80.50
16430.20	**METERING**				
0500	Outdoor wp meter sockets, 1 gang, 240v, 1 phase				
0510	Includes sealing ring, 100a	EA.	93.00	43.00	136
0520	150a	"	110	57.00	167
0530	200a	"	120	72.00	192

		UNIT	LABOR	MAT.	TOTAL
16470.10	**PANELBOARDS**				
1000	Indoor load center, 1 phase 240v main lug only				
1020	30a - 2 spaces	EA.	120	20.25	140
1030	100a - 8 spaces	"	150	64.00	214
1040	150a - 16 spaces	"	180	150	330
1050	200a - 24 spaces	"	210	350	560
1060	200a - 42 spaces	"	250	350	600
16490.10	**SWITCHES**				
4000	Photo electric switches				
4010	1000 watt				
4020	105-135v	EA.	44.75	33.50	78.25
4970	Dimmer switch and switch plate				
4990	600w	EA.	19.00	30.75	49.75
5171	Contractor grade wall switch 15a, 120v				
5172	Single pole	EA.	9.87	1.62	11.49
5173	Three way	"	12.25	2.97	15.22
5174	Four way	"	16.50	10.00	26.50
16510.05	**INTERIOR LIGHTING**				
0005	Recessed fluorescent fixtures, 2'x2'				
0010	2 lamp	EA.	44.75	63.00	108
0020	4 lamp	"	44.75	85.00	130
0205	Surface mounted incandescent fixtures				
0210	40w	EA.	41.25	75.00	116
0220	75w	"	41.25	82.00	123
0230	100w	"	41.25	97.00	138
0240	150w	"	41.25	110	151
0289	Recessed incandescent fixtures				
0290	40w	EA.	93.00	130	223
0300	75w	"	93.00	140	233
0310	100w	"	93.00	150	243
0320	150w	"	93.00	160	253
0395	Light track single circuit				
0400	2'	EA.	30.75	32.00	62.75
0410	4'	"	30.75	38.00	68.75
0420	8'	"	62.00	52.00	114
0430	12'	"	93.00	73.00	166
16510.10	**LIGHTING INDUSTRIAL**				
0500	Strip fluorescent				
0510	4'				
0520	1 lamp	EA.	41.25	43.25	84.50
0540	2 lamps	"	41.25	53.00	94.25
0550	8'				
0560	1 lamp	EA.	44.75	48.25	93.00
0580	2 lamps	"	55.00	88.00	143
1000	Parabolic troffer, 2'x2'				
1020	With 2 "U" lamps	EA.	62.00	120	182
1060	With 3 "U" lamps	"	71.00	140	211
1080	2'x4'				
1100	With 2 40w lamps	EA.	71.00	110	181
1120	With 3 40w lamps	"	82.00	110	192
1140	With 4 40w lamps	"	82.00	110	192
3120	High pressure sodium, hi-bay open				

		UNIT	LABOR	MAT.	TOTAL
16510.10	**LIGHTING INDUSTRIAL, Cont'd...**				
3140	400w	EA.	110	430	540
3160	1000w	"	150	750	900
3170	Enclosed				
3180	400w	EA.	150	700	850
3200	1000w	"	180	980	1,160
3210	Metal halide hi-bay, open				
3220	400w	EA.	110	270	380
3240	1000w	"	150	460	610
3250	Enclosed				
3260	400w	EA.	150	610	760
3280	1000w	"	180	580	760
3590	Metal halide, low bay, pendant mounted				
3600	175w	EA.	82.00	350	432
3620	250w	"	99.00	480	579
3660	400w	"	140	520	660
16710.10	**COMMUNICATIONS COMPONENTS**				
0020	Port desktop switch unit-4	EA.			77.00
0030	(USB)-4	"			200
0040	8	"			330
0050	16	"			550
0060	32	"			3,520
0070	Port console unit-8	"			2,200
0080	16	"			2,750
0090	Cat 5EJack and RJ45 coupler	"			3.74
1000	Quick-Port and voice grade	"			4.23
1010	Trac-Jack category 5E and connectors	"			4.21
1020	Snap-In connector	"			5.77
1030	Fast ethernet media converter	"			140
1040	Gigabit media converter	"			390
1050	With link fault signaling	"			130
1060	Gigabit switching media converter	"			550
1070	16-Bay media chassis	"			670
1080	Fiber-optic cable, Single-Mode, 50/125 microns, 10' length	"			22.00
1090	Multi-Mode, 62.5/125 microns, 10' length	"			25.25
2000	Fiber-Optic connectors, Unicam	"			17.00
2010	Fast-cam	"			17.50
2020	Adhesive style	"			6.32
2030	Threat-lock	"			10.75
2040	Unicam, high performance, single-mode	"			22.00
2050	multi-mode	"			19.75
2060	Network Cable, Cat5, solid PVC, 50'	"			16.00
2070	1000' Cat5, stranded PVC	"			190
2080	plenum PVC	"			220
2090	1000' Cat6, solid PVC	"			150
3000	USB Cables, 5 in 1 connector, male and female	"			19.75
3010	3 in 1 quick-connect, 4-pin or 6-pin	"			24.25
3020	Squid hub	"			33.00
0010	Key-stone jack module	"			3.30
0020	F-Type quick port	"			2.31
0030	BNC quick port	"			5.50
0040	RCA jack bulkhead connector	"			4.40

		UNIT	LABOR	MAT.	TOTAL
16760.00	**COMMUNICATIONS COMPONENTS, Cont'd...**				
0050	Quick-Port	EA.			5.72
0060	Snap-in	"			8.80
0070	HDMI wall plate and jack	"			10.50
0080	Voice/Data adapter	"			3.85
16760.10	**AUDIO/VIDEO CABLES**				
0010	Monster cable, 24k. gold plated	EA.			27.50
0020	DVI-HDMI	"			11.75
0030	Satellite/Video	"			3.96
0040	Digital/Optical	"			12.75
0050	Video/RCA	"			7.42
0060	S-Video	"			8.25
0070	Monster/S-video	"			44.00
0080	Speaker, clear jacket	FT.			0.93
0090	Speaker, high performance	"			2.20
0100	Flex-Premiere, oxygen free	"			0.33
16850.10	**ELECTRIC HEATING**				
1000	Baseboard heater				
1020	2', 375w	EA.	62.00	41.75	104
1040	3', 500w	"	62.00	49.50	112
1060	4', 750w	"	71.00	55.00	126
1100	5', 935w	"	82.00	78.00	160
1120	6', 1125w	"	99.00	92.00	191
1140	7', 1310w	"	110	100	210
1160	8', 1500w	"	120	120	240
1180	9', 1680w	"	140	130	270
1200	10', 1875w	"	140	180	320
1210	Unit heater, wall mounted				
1215	750w	EA.	99.00	140	239
1220	1500w	"	100	220	320
16910.40	**CONTROL CABLE**				
0980	Control cable, 600v, #14 THWN, PVC jacket				
1000	2 wire	L.F.	0.49	0.31	0.80
1020	4 wire	"	0.61	0.50	1.11

Average Square Foot Costs
2009

Housing (with Wood Frame)

Square Foot Costs do not include Garages, Land, Furnishings, Equipment, Landscaping, Financing, or Architect's Fee. Total Costs, except General Contractor's Fee, are included in each division. See Page 2 for Metropolitan Cost Variation Modifier. See Division 13 for Metal Buildings, Greenhouses, and Air-Supported Structures.

Division	Houses - 3 Bedrooms with Basement			Houses - 3 Bedrooms without Basement		
	Low Rent	Tract or Project	Custom	Low Rent	Tract or Project	Custom
1. General Conditions	$4.78	$4.88	$5.30	$6.20	$6.41	$6.83
2. Site Work	3.75	4.28	4.60	6.31	6.61	6.85
3. Concrete	4.33	4.55	4.84	7.85	8.05	8.92
4. Masonry	5.41	5.62	7.47	7.69	8.01	10.62
5. Steel	1.75	1.75	1.89	.86	.85	.87
6. Carpentry	18.14	17.81	20.07	26.60	26.38	28.69
7. Moisture & Therm. Prot.	4.55	4.55	4.67	6.33	6.56	7.13
8. Doors, Windows & Glass	9.20	8.88	9.26	11.34	11.66	12.23
9. Finishes	8.39	9.16	11.17	12.43	12.62	17.88
10. Specialties	.85	.87	.95	.93	.94	1.25
11. Equipment (Cabinets)	3.07	3.13	4.74	5.51	5.46	8.23
12. Furnishings	.63	.63	.68	.97	.97	.97
13. Special Construction	-	-	-	-	-	-
14. Conveying	-	-	-	-	-	-
15. Mechanical: Plumbing	7.28	7.34	8.62	10.77	10.72	13.33
Heating	4.41	4.57	4.75	6.21	6.55	9.24
Air Cond.	-	-	-	-	-	-
16. Electrical	4.20	4.20	5.06	6.25	6.25	7.41
SUB TOTAL	$80.74	$82.20	$94.07	$116.25	$118.03	$140.45
CONTRACTOR'S SERVICES	6.95	7.14	8.14	10.08	10.37	13.55
TOTAL AVERAGE Sq.Ft. Cost	$87.69	$89.34	$102.21	$126.33	$128.41	$153.99

	Unfinished Basement, Storage and Utility Areas included in Sq.Ft. Costs:			No Basement Utilities In Finished Area		
Average Cost/ Unit	$175,376	$178,679	$255,530	$151,599	$154,089	$230,990
Average Sq.Ft. Size/Unit (includes basement)	2,000	2,000	2,500	1,200	1,200	1,500
Add Architect's Fee	$ 7,023	$7,865	$17,978	$6,067	$6,854	$13,483

	COST	
	Sq. Ft.	Ea. Unit
Add for Finishing Basement Space For Bedrooms or Amusement Rooms, 12' x 20'	$33.08	$7,938
Add for Single Garage (no Interior Finish), 12' x 20'	$31.50	$7,560
Add for Double Garage (no Interior Finish), 20 x 20'	$33.08	$7,938
Add (or Deduct) for Bedrooms, 12' x 12'	$47.04	$6,774
Add per Bathroom with 2 Fixtures, 6' x 8'	$147.42	$7,076
Add per Bathroom with 3 Fixtures, 8' x 8'	$147.42	$9,435
Add for Fireplace	-	$7,350
Add for Central Air Conditioning	-	$6,090
Add for Drain Tiling	-	$3,318
Add for Well	-	$7,560
Add for Septic System	-	$7,560

Average Square Foot Costs
2009

Apartments, Hotels

Square Foot Costs do not include Land, Furnishings, Equipment, Landscaping and Financing.

		Apartments			Motels & Hotels		
	Division	2 & 3 Story	2 & 3 Story	High Rise	High Rise	1 & 2 Story	High Rise
		(1)a	(1)b	(2)	(3)	(1)b	(2)
1.	General Conditions	$ 4.52	$ 4.78	$ 6.83	$ 7.30	$ 4.94	$ 6.30
2.	Site Work	3.22	3.17	4.57	4.49	6.63	5.14
3.	Concrete	2.55	2.54	24.29	16.39	2.97	24.88
4.	Masonry	1.86	11.77	3.05	2.92	11.71	15.16
5.	Steel	1.66	1.67	2.65	2.76	2.01	2.53
6.	Carpentry and Millwork	19.04	15.97	6.16	5.71	15.99	4.96
7.	Moisture Protection	4.17	4.41	2.25	2.20	4.07	2.25
8.	Doors, Windows and Glass	6.53	7.10	8.08	7.84	8.14	9.72
9.	Finishes	12.79	12.82	13.13	12.37	12.79	17.60
10.	Specialties	.95	.96	1.00	.99	1.17	1.19
11.	Equipment	1.95	2.58	1.99	1.95	3.23	3.45
12.	Furnishings	.97	1.05	1.11	1.13	2.26	1.72
13.	Special Construction	-	-	-	-	-	-
14.	Conveying	-	2.73	4.12	4.25	-	4.78
15.	Mechanical:						
	Plumbing	8.62	8.62	8.46	8.31	9.80	13.22
	Heating and Ventilation	6.61	7.15	7.92	7.60	6.87	8.92
	Air Conditioning	2.35	2.35	2.46	2.41	2.91	5.78
	Sprinklers	-	-	2.62	2.32	2.43	2.48
16.	Electrical	9.07	9.02	10.36	10.75	12.10	12.80
	SUB TOTAL	$86.85	$98.68	$111.04	$101.70	$110.02	$142.86
	CONTRACTOR'S SERVICES	8.35	9.53	10.71	9.88	10.43	13.86
	Construction Sq.Ft. Cost	$95.20	$108.21	$121.75	$111.58	$120.45	$156.72
	Add for Architect's Fee	6.49	7.42	7.42	7.71	8.11	10.81
	TOTAL Sq.Ft. Cost	$101.69	$115.63	$129.17	$119.29	$128.55	$167.54
	Average Cost per Unit without Architect's Fee	$90,440	$102,802	$109,573	$89,265	$78,290	$101,871
	Average Cost per Unit with Architect's Fee	$96,602	$109,851	$116,251	$95,430	$83,561	$108,899

Add for Porches – Wood - $12.00 SqFt
Add for Interior Garages - $11.40 SqFt

	2 & 3 Story	2 & 3 Story	High Rise	High Rise	1 & 2 Story	High Rise
Average Sq.Ft. Size per Unit	950	950	900	800	650	650
Average Sq.Ft. Living Space	750	750	750	600	450	450

(1)a Wall Bearing Wood Frame & Wood or Aluminum Façade
(1)b Wall Bearing Masonry & Wood Joists & Brick Façade
(2) Concrete Frame
(3) Post Tensioned Concrete Slabs, Sheer Walls, Window Wall Exteriors, and Drywall Partitions.

Average Square Foot Costs
2009

Commercial

Square Foot Costs do <u>not</u> include Land, Furnishings, and Financing.

	Division	Remodeling Office Interior	Office 1-Story (1)	Office 2-Story (2)	Office Up To 5-Story (3)	High Rise (City) Above 5-Story (4)	Park-ing Ramp (3)
1.	General Conditions	$ 2.89	$ 4.67	$ 4.67	6.93	8.30	5.15
2.	Site Work & Demolition	1.89	2.71	2.69	2.63	2.46	1.77
3.	Concrete	-	7.81	12.13	13.38	9.10	29.91
4.	Masonry	-	11.82	11.24	16.96	4.21	1.70
5.	Steel	-	3.88	12.83	4.49	3.16	2.44
6.	Carpentry and Millwork	9.35	4.32	4.03	6.27	7.08	1.42
7.	Moisture Protection	-	7.76	3.87	2.57	3.26	3.61
8.	Doors, Windows and Glass	1.93	7.49	7.43	15.11	28.65	1.55
9.	Finishes & Sheet Rock	13.30	14.36	14.00	16.16	16.71	2.16
10.	Specialties	-	1.91	1.91	1.96	1.96	-
11.	Equipment	-	6.02	5.33	4.40	5.99	-
12.	Furnishings	-	1.24	1.47	1.63	1.76	-
13.	Special Construction	-	-	-	-	-	-
14.	Conveying	-	-	3.41	5.88	6.41	3.26
15.	Mechanical:						
	Plumbing	4.52	6.83	7.34	8.44	8.79	2.96
	Heating and Ventilation	2.03	7.66	7.49	9.68	9.74	-
	Air Conditioning	1.97	7.62	7.84	9.31	9.29	-
	Sprinklers	-	2.44	2.43	2.48	2.51	-
16.	Electrical	8.74	10.72	11.80	12.99	15.23	3.38
	SUB TOTAL	$46.62	$109.25	$121.92	$141.28	$ 144.60	$59.28
	CONTRACTOR'S SERVICES	4.46	9.16	11.76	12.31	12.29	5.68
	Construction Sq.Ft. Cost	$51.08	$118.41	$133.68	$153.58	$156.89	$64.96
	Add for Architect's Fee	4.95	8.75	9.14	11.02	12.34	5.04
	TOTAL Sq.Ft. Costs	$56.03	$127.15	$142.81	$164.61	$169.22	$70.00
	Deduct for Tenant Areas Unfinished	-	-	23.10	26.25	30.45	-
	Per Car Average	-	-	-	-	-	$18,375
	Add: Masonry Enclosed Ramp	-	-	-	-	-	7.04
	Add: Heated Ramp	-	-	-	-	-	7.04
	Add: Sprinklers for Closed Ramps	-	-	-	-	-	2.73
	Deduct for Precast Concrete Ramp	-	-	-	-	-	9.08

(1) Wall Bearing Masonry and Steel Joists and Decks – No Basement
(2) Wall Bearing Masonry and Precast Concrete Decks – No Basement
(3) Concrete Frame
(4) Steel Frame and Window Wall Façade

Average Square Foot Costs
2009

Commercial and Industrial

Square Foot Costs do not include Land, Furnishings, Landscaping, and Financing.
See Division 13 for Metal Buildings

	Division	Store 1-Story (1)	Store 2-Story (2)	Produc- tion 1-Story (1)	Ware- house 1-Story (1)	Ware- house and 1-Story Office (1)	Ware- house and 2-Story Office (3)
1.	General Conditions	$ 4.31	$ 4.31	$ 4.15	$ 3.71	$ 3.83	$ 3.49
2.	Site Work & Demolition	2.40	1.87	2.17	2.12	2.51	2.03
3.	Concrete	6.85	13.24	6.96	6.74	7.47	11.45
4.	Masonry	5.88	10.02	7.82	7.84	8.64	4.77
5.	Steel	12.31	6.44	12.31	12.04	13.11	9.89
6.	Carpentry	3.27	3.33	2.13	2.15	2.89	2.51
7.	Moisture Protection	5.46	3.17	5.99	5.78	5.80	2.89
8.	Doors, Windows and Glass	6.31	5.72	4.74	2.57	5.17	4.52
9.	Finishes	7.28	6.85	6.25	2.22	3.89	6.00
10.	Specialties	1.78	1.80	1.38	1.30	1.30	1.30
11.	Equipment	1.70	1.94	1.82	-	1.04	1.17
12.	Furnishings	-	-	-	-	-	-
13.	Special Construction	-	-	-	-	-	-
14.	Conveying	-	3.94	-	-	-	-
15.	Mechanical:						
	Plumbing	5.38	4.61	5.63	3.25	4.09	3.81
	Heating and Ventilation	4.96	4.92	3.82	3.94	5.33	6.13
	Air Conditioning	5.14	5.52	3.64	-	1.36	1.23
	Sprinklers	2.44	2.38	2.30	2.38	2.51	2.43
16.	Electrical	9.97	9.50	10.42	6.33	6.53	5.96
	SUB TOTAL	$85.42	$89.54	$81.52	$62.36	$75.47	$69.56
	CONTRACTOR'S SERVICES	8.19	8.61	7.85	6.00	7.28	6.70
	Construction Sq.Ft. Cost	$93.61	$98.15	$89.37	$68.36	$82.74	$76.26
	Add for Architect's Fees	6.36	6.69	6.11	4.66	5.65	5.22
	TOTAL Sq.Ft. Cost	$99.97	$104.84	$95.48	$73.02	$88.39	$81.48

(1) Wall Bearing Masonry and Steel Joists and Decks – Brick Fronts
(2) Wall Bearing Masonry and Precast Concrete Decks – Brick Fronts
(3) Precast Concrete Wall Panels, Steel Joists, and Deck

Average Square Foot Costs
2009

Educational and Religious

Square Foot Costs do not include Land, Furnishings, Landscaping and Financing.

Division	Elementary School 1-Story (1)	High School 2-Story (1)	Vocational School 3-Story (1)	College Multi-Story (2)	Gym (3)	Dorm 3-Story (1)	Church (3)
1. General Conditions	$ 7.46	$ 7.25	$ 7.35	$ 7.25	$ 5.88	$ 6.56	7.04
2. Site Work	4.39	4.60	4.49	4.39	2.19	2.89	4.24
3. Concrete	6.96	7.38	5.35	24.93	7.38	5.56	6.74
4. Masonry	14.73	13.36	14.52	15.16	15.05	13.46	30.63
5. Steel	12.54	14.15	14.84	5.23	15.35	13.34	2.25
6. Carpentry	4.65	4.59	4.20	3.08	2.61	4.20	8.51
7. Moisture Protection	6.26	4.28	4.82	4.84	6.63	5.80	6.90
8. Doors-Windows-Glass	7.22	7.28	7.60	9.63	6.74	6.26	8.82
9. Finishes	12.86	10.68	11.99	10.52	9.05	12.32	13.03
10. Specialties	2.01	1.86	1.86	2.28	1.86	1.77	1.38
11. Equipment	5.14	5.19	7.63	6.63	10.01	6.31	9.14
12. Furnishings	-	1.07	1.31	1.37	-	1.31	5.36
13. Special Construction	-	-	-	-	-	-	-
14. Conveying	-	-	3.73	3.73	-	3.73	-
15. Mechanical:							
Plumbing	11.26	10.86	10.92	14.67	6.83	10.08	6.97
Heating-Ventilation	14.07	13.80	15.62	14.23	8.99	9.84	9.84
Air Conditioning	-	-	-	-	-	-	7.49
Sprinklers	2.48	2.48	2.51	2.54	1.30	2.54	
16. Electrical	16.24	16.91	17.47	17.75	11.09	12.66	11.11
SUB TOTAL	$128.26	$125.74	$136.20	$148.21	$110.96	$118.63	139.44
CONTRACTOR'S SERVICES	12.25	14.05	13.02	14.14	10.61	11.26	13.42
Constr. Sq.Ft. Cost	$140.52	$139.79	$149.22	$162.35	$121.56	$129.88	152.86
Add Architect's Fees	9.46	9.29	10.11	11.13	8.25	8.75	10.43
TOTAL Sq.Ft. Cost	$149.97	$149.08	$159.33	$173.48	$129.81	$138.63	$163.29

(1) Wall Bearing Masonry and Steel Joists and Decks
(2) Concrete Frame and Masonry Façade
(3) Masonry Walls and Wood Roof

Average Square Foot Costs
2009

Medical and Institutional

Square Foot Costs do not include Land, Landscaping, Furnishings, and Financing.

		Clinic 1-Story (1)	Doctors Office 1-Story (1)	Hospital 1-Story (1)	Hospital Multi-Story (2)	Nursing Home 1-Story (1)	Housing for Elderly 1-Story (1)
1.	General Conditions	$ 7.88	$ 7.77	$ 10.34	$ 10.45	$ 9.40	$ 8.72
2.	Site Work	2.70	2.64	4.24	4.28	4.44	4.60
3.	Concrete	8.39	8.32	8.75	26.59	9.10	9.13
4.	Masonry	13.57	12.35	22.47	15.71	16.03	11.77
5.	Steel	10.82	10.70	13.17	2.92	2.93	2.44
6.	Carpentry	8.01	7.90	7.00	4.57	4.93	4.65
7.	Moisture Protection	5.99	5.94	6.45	2.68	6.44	6.40
8.	Doors, Windows and Glass	9.12	9.52	12.12	8.04	10.51	8.40
9.	Finishes	12.99	12.04	16.48	15.91	15.98	13.32
10.	Specialties	1.70	1.70	1.83	1.75	1.75	1.72
11.	Equipment	9.16	8.69	2.16	13.78	7.01	3.11
12.	Furnishings	1.37	1.37	1.45	1.50	1.28	2.72
13.	Special Construction	1.01	.99	2.01	2.01	1.00	1.01
14.	Conveying	-	-	-	7.56	-	6.93
15.	Mechanical:						
	Plumbing	14.06	15.14	23.13	22.53	15.34	14.84
	Heating and Ventilation	11.56	10.20	15.04	15.52	7.78	6.26
	Air Conditioning	8.88	8.62	15.41	13.40	8.97	3.37
	Sprinklers	2.51	2.48	2.51	2.53	2.86	2.51
16.	Electrical	18.54	15.12	28.11	26.10	17.05	15.01
	SUB TOTAL	$148.22	$141.49	$192.68	$197.82	$142.78	$126.88
	CONTRACTOR'S SERVICES	14.26	13.78	18.46	19.09	13.73	12.24
	Construction Sq.Ft. Cost	$162.48	$155.27	$211.14	$216.90	$156.51	$139.12
	Add for Architect's Fees	11.06	10.71	14.35	14.84	10.67	9.51
	TOTAL Sq.Ft. Cost	$173.53	$165.97	$225.50	$231.74	$167.19	$148.63

Average Cost to Remodel Approximately 50% Cost of New

---

(1) Wall Bearing Masonry and Steel Joists and Decks
(2) Concrete Frame and Masonry Façade

Average Square Foot Costs
2009

Government and Finance

Square Foot Costs do <u>not</u> include Land, Landscaping, Furnishings, and Financing.

		Bank 1-Story (1)	Bank & Office Bldg High- Rise (3)	Govt. Office Bldg Multi- Story (2)	Post Office 1-Story (1)	Court House 3-Story (2)	Jail and Prison (2)
1.	General Conditions	$ 8.51	$ 9.77	$ 9.77	$ 7.93	$ 9.66	$ 11.81
2.	Site Work	5.19	4.71	3.98	3.96	3.60	4.77
3.	Concrete	11.56	27.50	29.75	12.68	25.36	34.67
4.	Masonry	16.32	11.40	19.03	11.41	12.21	20.74
5.	Steel	6.04	2.70	3.25	4.72	2.82	6.73
6.	Carpentry	5.42	4.91	5.04	4.42	3.11	5.58
7.	Moisture Protection	4.41	4.95	5.24	4.46	5.51	6.26
8.	Doors, Windows and Glass	11.81	27.55	12.07	9.20	8.67	11.34
9.	Finishes	17.55	19.62	20.82	17.33	16.90	14.33
10.	Specialties	1.80	1.80	3.48	1.88	3.07	7.42
11.	Equipment	14.58	9.70	9.59	12.77	12.22	14.26
12.	Furnishings	1.62	1.47	-	2.15	-	-
13.	Special Construction	-	-	-	-	-	-
14.	Conveying	-	3.10	3.89	2.00	2.84	3.83
15.	Mechanical						
	Plumbing	8.62	10.08	10.00	10.70	13.72	23.80
	Heating and Ventilation	7.06	8.75	9.39	6.74	10.03	17.85
	Air Conditioning	5.67	6.44	8.17	8.13	14.32	14.84
	Sprinklers	2.65	2.54	2.48	2.54	2.54	2.54
16.	Electrical	17.25	20.83	21.50	16.46	20.50	31.79
	SUB TOTAL	$146.05	$177.81	$177.46	$139.47	$ 167.06	$232.56
	CONTRACTOR'S SERVICES	14.22	17.29	17.29	13.57	16.24	22.68
	Construction Sq.Ft. Cost	$160.27	$195.11	$194.75	$153.04	$183.30	$255.24
	Add for Architect's Fees	11.05	13.44	13.44	10.98	12.62	17.63
	TOTAL Sq.Ft. Cost	$171.31	$208.55	$208.19	$164.02	$195.92	$272.87

---

(1) Wall Bearing Masonry and Precast Concrete Decks
(2) Concrete Frame and Masonry Facade
(3) Concrete Frame and Window Wall

UNITED STATES INFLATION RATES
July 1, 1970 to June 30, 2008
AC&E PUBLISHING CO. INDEX

Start July 1

YEAR	AVERAGE %	LABOR %	MATERIAL %	CUMULATIVE %
1970-71	8	8	5	0
1971-72	11	15	9	11
1972-73	7	5	10	18
1973-74	13	9	17	31
1974-75	13	10	16	44
1975-76	8	7	15	52
1976-77	10	5	15	62
1977-78	10	5	13	72
1978-79	7	6	8	79
1979-81	10	5	13	89
1980-81	10	11	9	99
1981-82	10	11	10	109
1982-83	5	9	3	114
1983-84	4	5	3	118
1984-85	3	3	3	121
1985-86	3	2	3	124
1986-87	3	3	3	127
1987-88	3	3	4	130
1988-89	4	3	4	134
1989-90	3	3	3	137
1990-91	3	3	4	140
1991-92	4	4	4	144
1992-93	4	3	5	148
1993-94	4	3	5	152
1994-95	4	3	6	156
1995-96	4	3	6	160
1996-97	3	3	3	163
1997-98	4	3	4	167
1998-99	3	3	2	170
1999-2000	3	2	3	173
2000-2001	4	4	4	177
2001-2002	3	4	3	180
2002-2003	4	5	3	184
2003-2004	3	5	2	187
2004-2005	8	4	10	195
2005-2006	8	6	9	203
2006-2007	8	4	11	212
2007-2008	3	4	3	215
2008-2009	4	3	5	219

-A-

ABOVE GROUND TANK 02-12
ABRASIVE SURFACE 05-6
SURFACE TILE 09-5
ACCESS CONTROL 10-5
ACCORDION FOLDING DOORS 08-7
ACID-PROOF COUNTER 12-2
ACOUSTICAL BLOCK 04-7
PANEL 09-5
TILE 09-6
ACRYLIC CARPET 09-8
AD PLYWOOD 06-11
ADHESIVE-BED TILE 09-4
ADJUSTABLE SHELF 06-10
ADMIXTURE 03-8
AIR COMPRESSOR 01-4
ENTRAINING AGENT 03-8
TOOL 01-5
ALUMINUM DOOR 08-13
DOWNSPOUT 07-7
FLASHING 07-7
GUTTER 07-7
LOUVER 05-7
PLAQUE 10-4
RAILING 05-7
ROOF 07-6
SHINGLE 07-3
SIDING PANEL 07-4
STOREFRONT 08-8
THRESHOLD 08-12
TREAD 05-6
ANCHOR 05-4
BOLT 05-4
BRICK 04-3
DOVETAIL 04-3
SLOT 04-3
ANGLE STEEL 04-9
ANTI-SIPHON BREAKER 15-4
APRON 06-9
ARCHITECTURAL FEE 01-2
ARMOR PLATE 08-11
ASH RECEIVER 10-6
ASHLAR VENEER 04-11
ASPHALT DAMPPROOFING 07-2
EXPANSION JOINT 03-8
AUTOMOTIVE HOIST 14-3
AWNING WINDOW 08-10

-B-

BACK SPLASH 06-11
BACKFILL 02-7
HAND 02-8
BACKHOE 01-5
BACKHOE/LOADER 01-6
BACK-UP BLOCK 04-7
BRICK 04-4
BAKE OVEN 11-3
BALANCE LABORATORY 11-6
BALUSTER 06-10
BAND STAGE 11-2
BANK RUN GRAVEL 02-8
BAR REINFORCING 04-2
BARRICADE 01-3
BARRIER VAPOR 07-2
BASE CABINET 06-11, 12-2
COLONIAL 06-9
COLUMN 04-10
FLASHING 07-7
GRANITE 04-11
MANHOLE 02-10
BASEBOARD HEATER 16-8
BATCH DISPOSAL 15-5
BATH FAUCET 15-5
BATHROOM LOCK 08-11
BATTEN SIDING 07-5
BEAD MOLDING 06-9
PARTING 06-10
PLASTER 09-3, 09-4
BEAM 05-4
BOND 03-9, 04-2, 04-8
FURRING 05-4, 09-2
GRADE 03-10
BEAM PLASTER 09-3
REINFORCING 03-7
BENCH FLOOR MOUNTED 10-5
BEVELED SIDING 07-5
BIBB HOSE 15-3
BICYCLE EXERCISE 11-6
BI-FOLD DOOR 08-4
BIN ICE STORAGE 11-5
BINDER COURSE 02-10
BI-PASSING DOOR 08-4
BIRCH DOOR 08-3
BIRDSCREEN 13-2
BITUMINOUS MEMBRANE 07-2
SIDEWALK 02-10
BLOCK CONCRETE 03-11, 04-6
DEMOLITION 02-2
GLASS 04-10
GRANITE 04-6
GROUT 04-2
REMOVAL 02-4
BLOOD ANALYZER 11-6
REFRIGERATOR 11-6
BLOWN-IN INSULATION 07-3
BLUESTONE 04-6
BOARD DIVING 13-3
RAFTER 06-5
RIDGE 06-4
SIDING 07-5
BOLTED STEEL 05-4
BOND BEAM 04-2, 04-8
BEAM PLACEMENT 03-9
BONDERIZED FLAGPOLE 10-3
BOOKKEEPER 01-2
BOTTLE COOLER 11-5
BOTTOM BEAM 03-2
PIVOT HINGE 08-10
PLATE 06-6
BOX JUNCTION 16-5
RECEIVING 10-6
BREAKER CIRCUIT 16-2
VACUUM 15-4
BRICK ANCHOR 04-3
CHIMNEY 04-5
FIRE RATED 04-5
FIREWALL 04-5
MANHOLE 02-11
PAVER 04-6
REMOVAL 02-3, 02-4
BROILER KITCHEN 11-3
BRONZE PLAQUE 10-4
RAILING 05-7
BROOM FINISH 03-8
BUFF BRICK 04-4
BUILDING METAL 13-2
PAPER 07-2
BUILT-IN OVEN 11-5
BUILT-UP PLAQUE 10-4
ROOF 07-6
ROOF REMOVAL 02-3
BULKHEAD FORMWORK 03-5
BULLDOZER 01-6
BULLETIN BOARD 10-3
BULLNOSE GLAZED 04-8
BURLAP CONCRETE 03-9
RUB 03-9
BUS DUCT 16-2
BUTT HINGE 08-10
BX CABLE 16-4

-C-

CABINET BASE 06-11
KITCHEN 12-2
LABORATORY 11-6
CABLE BX 16-4
THHN-THWN 16-4
CAGE LADDER 05-6
CANT STRIP 06-5, 07-6
CAP PILE 03-4, 03-10
PILE REINFORCING 03-7
CARBORUNDUM 03-9
CARD ACCESS CONTROL 10-5
CARPET CLEAN 09-8
TILE 09-8
CARRIER CHANNEL 09-6
CARVED DOOR 08-4
CASE REFRIGERATED 11-4
CASED BORING 02-2
CASEMENT WINDOW 08-8, 08-9
CASING 06-9
CASING BEAD 09-4
RANCH 06-9
TRIM 06-10
CAST BOX 16-5
IRON STRAINER 15-4
IRON TREAD 05-5
CAVITY WALL 04-2, 04-4
WALL ANCHOR 04-3
CDX 06-6
CEDAR DECKING 06-7
SHINGLE 07-4
SIDING 07-5
CEILING DIFFUSER 15-9
FURRING 06-3, 09-2
HEATER 11-5
INSULATION 07-2
JOIST 06-2
REMOVAL 02-3
CEMENT FIBER 03-12
FIBER BOARD 03-12
PLASTER 09-3
CENTRIFUGE LABORATORY 11-6
CHAIN HOIST DOOR 08-7
LINK 02-5
LINK FENCES 09-9
CHAIR BAND 11-2
HYDROTHERAPY 11-6
LIFEGUARD 13-3
RAIL 06-9
CHAMFER STRIP 03-5
CHANNEL FURRING 05-4, 09-2
SLAB 03-12
CHECK VALVE 15-4
CHEMICAL EXTINGUISHER 10-5
CHIMNEY BRICK 04-5
CHUTE MAIL 10-6
CIRCUIT BREAKER 16-2
CLAY BRICK FLOOR 09-7
PIPE 02-12, 02-13
TILE FLOOR 04-10
CLEARING TREE 02-5
CLEFT NATURAL 04-6
CLOCK-TIMER HOSPITAL 11-8
CLOSET DOOR 08-4
POLE 06-9
CLOTH FACED FIBERGLASS 09-6
CMU 04-6
GROUT 04-2
CO2 EXTINGUISHER 10-5
COFFEE URN 11-3
COILING DOOR 08-6
COLONIAL BASE 06-9
MOLDING 06-10
COLOR GROUP TILE 09-5
COLUMN 05-4
BASE 04-10
COVER 08-13
FOOTING 03-4
FURRING 05-4
PIER 03-4
PLASTER 09-3
TIMBER 06-7
COMMERCIAL DOOR 08-8
COMMON BRICK 04-4
COMPACT BASE 02-8
BORROW 02-7
KITCHEN 11-3
COMPACTOR INDUSTRIAL 11-3
RESIDENTIAL 11-5
COMPARTMENT SHOWER 10-2
COMPRESSION FITTING 16-2
COMPRESSOR 01-4
CONCEALED Z BAR 09-6
CONCRETE BLOCK 09-9
BLOCK REMOVAL 02-4
CUTTING 02-4
DEMOLITION 02-2
DOWEL 03-7
FILLED PILE 02-9

CONCRETE HARDENER 03-9
  MANHOLE 02-10
  PILING 02-9
  PLANK 03-12
  PUMP 03-9
  RECEPTOR 10-2
  REINFORCEMENT 03-6
  REMOVAL 02-4
CONCRETE SLEEPER 06-5
  TESTING 01-3
CONDUCTIVE FLOOR 09-7
CONDUCTOR COPPER 16-4
CONDUIT EMT 16-2
  FLEXIBLE 16-2
  GALVANIZED 16-2
  PVC 16-3
  STEEL 16-2
CONNECTOR COMPRESSION 16-2
CONTINUOUS DISPOSAL 15-5
  FOOTING 03-10
CONTROL JOINT 04-4
CONVECTION OVEN 11-3
COOLER BOTTLE 11-5
COPING CLAY 04-10
  LIMESTONE 04-11
COPPER DOWNSPOUT 07-7
  FLASHING 07-7
  PLAQUE 10-4
CORNER BEAD 09-3, 09-4
  CABINET 12-2
CORNICE MOLDING 06-10
CORRUGATED METAL PIPE 02-12
  PANEL 07-5
  ROOF 07-6
COUNTER DOOR 08-6
  FLASHING 07-7
  GRIDDLE 11-4
  LABORATORY 11-6
  LAVATORY 15-6
COUNTERTOP 06-11
COUNTER-TOP RANGE 11-5
COUPLING PLASTIC 02-11
COURSE BINDER 02-10
COVE BASE 09-7
  BASE GLAZED 04-8
  BASE TILE 09-5
  MOLDING 06-9
COVER POOL 13-3
CRACK REPAIR 03-13
CRANE 01-6
  MOBILIZATION 01-3
CRAWL SPACE INSULATION 07-2
CRAWLER CRANE 01-6
CREOSOTE 06-11
CREW TRAILER 01-3
CROWN MOLDING 06-9
CRUSHED MARBLE 04-6
CURB FORM 03-3
  FORMWORK 03-6
  GRANITE 04-11
  TERRAZZO 09-5
CURING PAPER 03-9
CURTAIN WALL 08-13
CURVED CURB FORM 03-3
CUSTOM DOOR 08-6
CUT AND FILL 02-5
  STONE 04-10
  TREE 02-5
CUT-OUT 02-3
  COUNTERTOP 06-11
CYLINDER PILING 02-9

-D-

DAIRY PRODUCT 11-4
DARBY 03-8
DECKING WOOD 06-7
DECKS, WOOD, STAINED 09-10
DELICATESSEN CASE 11-4
DIAPHRAGM PUMP 01-5
DIMMER SWITCH 16-6
DIRECTORY BOARD 10-3
DISHWASHER 11-4
  RESIDENTIAL 11-5

DISPENSER TISSUE 10-7
DISPOSAL COMMERCIAL 11-4
  RESIDENTIAL 11-5
DIVIDER STRIP TERRAZZO 09-5
DIVING BOARD 13-3
DOCK LEVELER 11-2
DOMESTIC MARBLE 04-11
  WELL 02-12
DOOR ALUMINUM 08-13
  CHAIN HOIST 08-7
  CLOSER 08-11
DOOR FLUSH 08-4
  FRAME 08-2
  FRAME GROUT 04-2
  HOLDER 08-11
  MOTOR 08-6
  OVERHEAD 08-6
  PLATE 08-11
  REMOVAL 02-2
  REVOLVING 08-8
  ROLL UP 13-2
  SLIDING GLASS 08-7
  SOLID CORE 08-3
DOORS 09-8
  AND WINDOWS 09-9
DOORS, WOOD 09-10, 09-12
DOUBLE HUNG WINDOW 08-9
  OVEN 11-3
  WALLED TANK 02-12
DOUGLAS FIR 06-7
DOVETAIL ANCHOR SLOT 04-3
DOWEL REINFORCING 03-7
DOWNSPOUT 07-7
  ALUMINUM 07-7
DOZER EXCAVATION 02-5
DRAIN FIELD 02-13
  TILE 02-12
DRAINAGE UNDERSLAB 02-13
DRAPERY TRACK 12-2
DRILLED WELL 02-12
DROP PANEL 03-3
DROPCLOTHS 09-8
DRY RUB 03-9
  STONE WALL 04-10
  TYPE EXTINGUISHER 10-5
DRYWALL REMOVAL 02-2, 02-3
DUCT BUS 16-2
DUCTWORK METAL 15-8
DUMP TRUCK 01-5
DUMPSTER 02-4
DUTCH DOOR 08-2
DWV COPPER 15-2
  FITTING COPPER 15-2

-E-

EDGE FORM 03-4
  FORMWORK 03-6
  REPAIR 03-13
ELASTIC SHEET ROOF 07-6
ELASTOMERIC FLASHING 04-4
ELBOW CAST IRON 02-11
  PLASTIC 02-11
ELECTRIC ELEVATOR 14-2
  FURNACE 15-8
ELEVATED SLAB 03-5
ELEVATOR 14-2
ENGINEER 01-2
ENGRAVED PLAQUE 10-4
ENTRANCE DOOR 08-8
ENTRY LOCK 08-11
EPOXY CONCRETE 03-13
  GROUT 03-12
EQUIPMENT CURB 03-3
  MOBILIZATION 01-3
  PAD 03-10
  PAD REINFORCING 03-6
EXERCISE BICYCLE 11-6
EXHAUST HOOD 11-4
EXPANDED LATH 09-3
EXPANSION JOINT 03-8, 09-3
EXPOSED AGGREGATE 04-6
EXTENSION BOX 16-5
EXTERIOR DOOR 08-4, 08-5

EXTINGUISHER FIRE 10-5
EXTRACTOR WASHER 11-2

-F-

FABRIC CHAIN LINK 02-13
  FORM 03-13
FABRICATION METAL 05-6
FACING PANEL 04-11
  PANELS 04-10
FACTORY FINISH FLOOR 09-6
FAN EXHAUST 15-8
FASCIA BOARD 06-5
  PLASTER 09-3
FAUCET 15-5
FEE PROFESSIONAL 01-2
FENCE 02-13
  REUSE 02-5
  WOOD 02-13
FENCES, WOOD OR MASONRY 09-10
FIBER BOARD 03-12
  CANT 06-5
  FORM CONCRETE 03-3
  PANEL 09-6
FIBERBOARD SHEATHING 06-7
FIBERGLASS DIVING BOARD 13-3
  INSULATION 07-2
  PANEL 09-6
  TANK 02-12
FIBER-OPTIC CABLE 16-7
FIBROUS DAMPPROOFING 07-2
FIELD ENGINEER 01-2
FILL BORROW 02-6
  CAP 02-12
  GYPSUM 03-12
FILLED BLOCK 03-11, 04-7
FILLER JOINT 04-2
FILTER BARRIER 02-9
FINISH PLASTER 09-3
  SHOTCRETE 03-9
FINISHING SHEETROCK 09-4
FIR DECKING 06-7
  FLOOR 09-6
  SIDING 07-5
FIRE DOOR 08-6
  ESCAPE 05-7
  EXTINGUISHER 10-5
  RATED BRICK 04-5
  RATED DOOR 08-2
  RESISTANT SHEETROCK 09-4
  RETARDANT 06-11
  RETARDENT ROOF 07-4
FIREWALL BRICK 04-5
FIXED SASH 08-8
  TYPE LOUVER 05-7
  WINDOW 08-8
FIXTURE PLUMBING 15-4
FLAGSTONE PAVER 04-6
FLASHING 07-6
  MASONRY 04-4
FLAT APRON 06-9
  CONDUIT CABLE 16-2
FLOAT 03-8
FLOOR CLAY TILE 04-10
  CLEANOUT 15-3
  DRAIN 15-4
  FINISH 03-8
  FINISHING 09-6
  GYM 09-6
  HARDENER 03-9
  JOIST 06-2
  MASONRY 09-7
  MOUNTED BENCH 10-5
  PATCH 03-12
  REMOVAL 02-3
  SAFE 13-2
  SHEATHING 06-6
  SLAB 03-3
  TERRAZZO 09-5
  TILE 09-4, 09-5
FLOORING MARBLE 04-11
FLUID APPLIED FLOOR 09-7
FLUORESCENT FIXTURE 16-6
FLUSH DOOR 08-2, 08-4

INDEX

FLUSH PANEL DOOR 08-13
FLUTED COLUMN 04-10
    STEEL CASING........................... 02-9
FOAMGLASS CANT 07-6
FOIL BACKED INSULATION 07-3
FOOTING EXCAVATION 02-7
FOREMAN 01-2
FORM FABRIC 03-13
FORMBOARD 03-12
FORMICA 06-11
    PARTITION 10-2
FORMWORK OPENINGS 03-4
FOUNDATION BACKFILL 02-7
    BLOCK 04-7
    VENT 04-2
FRAME COVER 02-11
    DOOR 08-5
FRAMING ROOF 06-3
    SOFFIT 06-5
    TIMBER 06-7
    WALL 06-5
FRENCH DOOR 08-4
FRICTION PILING 02-9
FRONT-END LOADER 02-5
FROZEN FOOD CASE 11-5
FRYER KITCHEN 11-3
FURRING CEILING 09-2
    CHANNEL 05-4
    STEEL 09-2

-G-

GABLE TRUSS 06-8
GABLE-END RAFTER 06-3
GALVANIZED DUCTWORK.................. 15-8
    MESH 03-7
    METAL DOOR 08-2
    RAILING.................................... 05-6
    REGLET 03-6
    REINFORCEMENT 03-6
    SHINGLE 07-3
    WALL TIE 04-2
GAS FURNACE 15-8
    STERILIZER 11-7
GATE VALVE................................... 15-4
GEL EPOXY 03-13
GENERATOR 01-5
GFI BREAKER 16-5
GIRDER 05-4
GLASS BEAD 06-9
    CLOTH TILE 09-6
    DOOR 08-7
    ENCASED BOARD........................ 10-3
    FIBER BOARD............................ 03-12
    FRAMING 08-13
GLASSWARE WASHER 11-6
GLAZED BLOCK 04-8
    BRICK 04-4
    FLOOR 09-7
    TILE 04-5
    WALL TILE 09-4
GLOBE VALVE 15-4
GLUE LAM 06-8
GRAB BAR 10-6
GRADALL 02-7
GRADING 02-5
GRANITE BLOCK 04-6
    VENEER 04-11
GRAVEL MULCH.............................. 02-16
    ROOF 07-6
GRIDDLE KITCHEN 11-4
GRILLE METAL 05-7
GROUND FAULT............................... 16-5
    ROD 16-5
    SLAB 03-5
GROUT BEAM 04-2
    BLOCK........................03-11, 04-2, 04-7
    CRACK SEAL 03-13
    EPOXY 03-13
    MASONRY 04-2
    WALL 04-2
GRS CONDUIT 16-2
GUNITE.. 03-9
GUTTER 07-7

GUTTER COPPER 07-7
    STEEL 07-7
GYM FLOOR.................................... 09-6
    LOCKER 10-5
GYPSUM FILL 03-12
    MASONRY 04-8
    PLANK 03-12
    PLASTER 09-3
    SHEATHING 06-7

-H-

HALF ROUND MOLDING.................... 06-9
HALIDE FIXTURE............................. 16-7
HAND BUGGY 03-9
    DRYER...................................... 10-7
    EXCAVATION 02-7
    SEEDING 02-15
    SPLIT SHAKE 07-4
    SPREADING 02-14
HANDICAPPED PARTITION.............. 10-2
    PLUMBING 15-5
HANDLING RUBBISH 02-4
HARDBOARD DOOR 08-3
    UNDERLAYMENT 06-6
HARDENER CONCRETE 03-9
HASP ASSEMBLY............................ 08-11
HAULING RUBBISH 02-4
HEADER 06-6
    CEILING 06-2
HEATER 01-5
    ELECTRIC 11-5, 16-8
    UNIT 15-8
    WATER 15-7
HEAVY DUTY DOOR 08-2
    DUTY TERRAZZO 09-5
HERRINGBONE FLOOR..................... 09-7
HIGH PRESSURE SODIUM 16-6
    STRENGTH BLOCK 04-7
HIP RAFTER 06-4
HOLLOW BLOCK 04-6
    CLAY TILE 04-9
    COLUMN 06-12
    CORE DOOR 08-3, 08-4
    METAL DOOR 08-2
HOOD EXHAUST 11-4
    RANGE 11-5
HORIZONTAL DAMPER 15-9
    REINFORCING....................03-8, 04-2
    SIDING 07-5
    WINDOW 08-9
HOSPITAL EQUIPMENT.................... 11-6
HOT PLATE LABORATORY 11-6
    WATER DISPENSER 15-5
H-SECTION PILE 02-9
HUNG SLAB 03-3
HYDRAULIC ELEVATOR.................... 14-2
    EXCAVATOR....................02-5, 02-7

-I-

ICE CUBE MAKER 11-4
    MAKER 11-5
    STORAGE BIN 11-5
IMC CONDUIT 16-3
INCANDESCENT FIXTURE 16-6
INCINERATOR LABORATORY 11-6
INCUBATOR 11-7
    LABORATORY 11-6
INDUSTRIAL COMPACTOR 11-3
    WINDOW 08-8
INJECTION GROUT 03-13
INSULATED BLOCK......................... 04-7
    GLASS 08-12
    GLASS DOOR 08-7
    SIDING 07-5
INSULATION BLOWN-IN 07-3
    SPRAYED 07-3
INTERIOR SIGN 10-4
    WALL FORM 03-5
INTERMEDIATE CONDUIT 16-3
    PIVOT HINGE 08-10
IRON TREAD 05-6
IRREGULAR STONE 04-6

IRRIGATION LAWN........................... 02-13
ISOLATION MONITOR 11-8

-J-

JACK RAFTER................................. 06-4
JALOUSIE WINDOW 08-8
JOB BUILT FORM 03-2
    SCHEDULING 01-3
    TRAILER 01-3
JOINT CLIP 09-3
    EXPANSION 03-8
    FILLER 04-4
    REINFORCING 04-2
JOIST SISTER 06-2, 06-3
JUMBO BRICK 04-4
JUTE PADDING 09-8

-K-

K COPPER 15-2
KETTLE STEAM 11-4
KEYWAY FORM 03-5
KICK PLATE 05-6, 08-11
KILN DRIED 06-12
KING POST TRUSS 06-8
KITCHEN CABINET...............06-11, 12-2
    COMPACT 11-3
    FAUCET 15-5
    SINK 15-6

-L-

L TUBE 15-2
LADDER POOL 13-3
    REINFORCING 04-2
LAMINATED BEAM 06-8
    PLASTIC 06-11
    RAILING.................................... 05-7
LANDING STEEL 05-5
LANDSCAPE TIMBER....................... 02-16
LATCHSET 08-11
LATH EXPANDED 09-3
    STUCCO 09-3
LATTICE MOLDING.......................... 06-9
LAUAN DOOR 08-3
LAVATORY FAUCET 15-5
LEACHING PIT 02-13
LEAD LINED DOOR 08-2
LEVEL INDICATOR 02-12
LEVELER DOCK.............................. 11-2
LIFEGUARD CHAIR 13-3
LIFT AUTO 14-3
LIGHT GAGE FRAMING 05-5
    POOL 13-3
    STEEL 05-4
    SURGICAL 11-7
LIGHTING 16-6
LIGHTWEIGHT BLOCK..................... 04-7
LIMESTONE AGGREGATE 04-6
    PANEL 04-11
LIMING 02-15
LINEN CHUTE 14-3
LINER PANEL 13-3
LINOLEUM REMOVAL 02-3
LINTEL LIMESTONE 04-11
    STEEL 04-9
LIQUID CHALKBOARD 10-2
LOAD BEARING STUD...................... 09-2
    CENTER BREAKER 16-5
    CENTER INDOOR 16-6
LOADER 01-6
LOCKSET 08-11
LOUVER DOOR 08-4
    METAL 05-7

-M-

M TUBE 15-2
MAIL CHUTE 10-6
MANHOLE....................................... 02-10
MAP RAIL CHALKBOARD 10-2
MARBLE CHIP 02-15
    IMPORTED 04-10

MASKING PAPER AND TAP 09-8
MASONRY BASE 04-10
MASONRY CHIMNEY 04-5
    DEMOLITION.......................... 02-2
    FACING PANEL 04-10
    FENCE 02-5
    FURRING 06-3
    GLASS 04-10
    GYPSUM 04-8
    LIMESTONE 04-11
    PLAQUE 10-4
    PLASTER 04-10
    REINFORCING 03-8
    SLATE 04-11
MAT FORMWORK 03-4
    REINFORCING 03-7, 03-8
MEAT CASE 11-5
MECHANICAL JOINT PIPE 02-11
    SEEDING 02-15
MEDICAL EQUIPMENT 11-6
MEDICINE CABINET 10-7
MEMBRANE CURING 03-9
    WATERPROOFING 07-2
MESH SLAB 03-7
    TIE 04-4
    WIRE 03-11
METAL BASE 09-3
    BUILDING 13-2
METAL CLAD DOOR 08-6
    DOOR REMOVAL 02-2
    FABRICATION 05-5
    FENCE 02-13
    HALIDE 16-7
    LADDER 05-6
    LINTEL 04-9
    LOUVER 08-2
    PAN STAIR 05-5
    PLAQUE 10-4
    RAILING 05-6, 05-7
    ROOF REMOVAL 02-3
    SHELVING 10-6
    STAIR 05-5
    STUD 05-5
    TOILET PARTITION 10-2
    WINDOW REMOVAL 02-3
METALLIC HARDENER 03-9
METER SERUM 11-6
MICROSCOPE LABORATORY 11-6
MILL FRAMED STRUCTURE 06-7
MILLED WINDOW 08-10
MINERAL FIBER BOARD 03-12
    FIBER PANEL 09-6
    FIBER TILE 09-6
MIRROR 10-7
    PLATE 08-13
MODIFIER EPOXY 03-13
MONOLITHIC TERRAZZO 09-5
MOP SINK 15-7
MORTISE LOCKSET 08-11
MOSS PEAT 02-15
MOTORIZED CHECKROOM 11-2
MOVABLE LOUVER 05-7
MOVING SHRUB 02-14
    TREE............................. 02-14
MULCH 02-15
MULLION WINDOW 08-8

-N-

NAILER CEILING.......................... 06-2
NAMEPLATE PLAQUE 10-4
NAPKIN DISPENSER........................ 10-7
NARROW STILE DOOR...................... 08-13
NATURAL CLEFT SLATE 04-6
NEOPRENE FLASHING 07-6
    JOINT 03-8
    MEMBRANE 07-2
    ROOF 07-6
NET SAFETY 01-4
NM CABLE 16-4
NO-HUB ADAPTER COPPER 15-3
    PIPE............................. 15-2
NON-DRILLING ANCHOR 05-4
NON-METALLIC HARDENER................... 03-9

NORMAN BRICK 04-5
NYLON CARPET 09-8

-O-

OAK DOOR................................ 08-6
    FLOOR 09-6
OFFICE TRAILER 01-3
OGEE MOLDING 06-10
OIL FURNACE 15-8
    TANK............................. 02-12
ORGANIC BED TILE 09-5
OUTRIGGER FLAGPOLE 10-3
OVEN BAKE 11-3
    CONVECTION 11-3
OVERHEAD 01-2
    DOOR 08-6, 08-7, 13-2
OVERSIZED BRICK 04-4

-P-

PACKAGED HEATER 15-8
PAD EQUIPMENT 03-3
    REINFORCING 03-6
PADDING CARPET 09-8
PALM TREE 02-14
PAN METAL STAIR 05-5
    SLAB 03-3
PANEL ACOUSTICAL 09-5
    LIMESTONE 04-11
PANEL SIDING 07-5
PANIC DEVICE 08-11
PAPER BACKED INSULATION 07-2
    BACKED LATHE 09-3
    BUILDING 07-2
    CURING 03-9
PARABOLIC TROFFER 16-6
PARALLEL DAMPER 15-9
    STRAND BEAM 06-8
PARGING 04-10
PARTING BEAD 06-10
PARTITION CLAY TILE 04-9
    REMOVAL 02-2
    TOILET 10-2
PASSENGER ELEVATOR 14-2
PASS-THROUGH WASHER 11-2
PATCH CONCRETE 03-9
PATIO BLOCK 04-6
PAVEMENT CUTTING 02-4
PAVER GRANITE 04-6
    MARBLE 04-11
    STONE 04-6, 04-11
PAVING CONCRETE 02-10
PEAT MOSS 02-15
PEDESTRIAN BARRICADE 01-3
PENNSYLVANIA SLATE 07-4
PERFORATED CLAY PIPE 02-12
    PVC PIPE 02-13
PERFORMANCE BOND 01-6
PERLITE 03-12
    ROOF INSULATION 07-3
PHYSICAL THERAPY 11-6
PICKUP TRUCK 01-5
PICTURE WINDOW 08-8, 08-10
PILASTER........................03-4, 03-11
PILE REPAIR 03-13
    SPALL 03-13
PILING MOBILIZATION 01-4
PINE DECKING 06-7
    DOOR 08-6
    RISER 06-12
    SIDING 07-6
PIPE CLAY 02-12
    COLUMN 05-4
    COPPER 15-2
    FLASHING 07-6
    GALVANIZED 15-3
    NO-HUB 15-2
    PILE POINT 02-10
    RAILING 05-6
PIVOT HINGE 08-10
PLANT BED PREPARATION 02-15
PLAQUE BRONZE 10-4
    MASONRY 10-4

PLASTER MASONRY 04-10
    REMOVAL 02-3
PLASTIC FILTER FABRIC 02-13
    SHEET 07-2
    SIGN 10-5
PLATE 06-6
    ARMOR 08-11
    GLASS 08-12
    MIRROR 08-12
PLEXIGLASS 08-12
PLUG CONDUIT 16-5
PLYWOOD 06-6
    FINISH 06-11
    SHEATHING 06-6
    SHELVING 06-10
    STRUCTURAL 06-7
    UNDERLAYMENT 06-6
PNEUMATIC HOIST 14-3
    TOOL 01-5
POLE CLOSET 06-9
POLYETHYLENE VAPOR BARRIER 07-2
    WRAP 03-13
POLYSTYRENE INSULATION 07-3
POLYURETHANE FLOOR 09-6
POOL EQUIPMENT 13-3
PORCELAIN SINK 15-7
POROUS CONCRETE PIPE 02-12
PORTABLE STAGE 11-2
POSTS TREATED 06-7
POURED DECK 03-12
PRECAST LINTEL 04-9
    MANHOLE 02-10
    SEPTIC TANK 02-13
PREFINISHED SHELVING 06-11
PRIMED DOOR 08-4
PRIVACY FENCE 02-13
    LOCKSET 08-11
PRODUCE CASE 11-5
PROFESSIONAL FEE 01-2
PROFIT 01-2
PROJECTING SASH........................ 08-8
    WINDOW 08-9
PULL BOX 16-5
PUMP 01-5
PUMPED CONCRETE 03-9, 03-11
PUSH PLATE 08-11
PUTTYING 09-9
PVC CHAMFER STRIP 03-5
    CONDUIT 16-3
    CONTROL JOINT 04-4
    FORM LINER 03-5
    PIPE 02-11, 02-12, 02-13
    ROOF 07-6
    SEWER PIPE 02-13

-Q-

QUARTER ROUND.......................... 06-10

-R-

RACEWAY 16-3
RACK-TYPE DISHWASHER................... 11-4
RADIOLOGY EQUIPMENT 11-7
RADIUS CURB 04-11
    FORMWORK 03-6
RAFTER 06-3
    HIP 06-4
RAIL CHAIR 06-9
RAILING 06-10
    METAL 05-7
RANCH CASING 06-9
    MOLDING 06-10
RANDOM WIDTH FLOOR 09-6
RANGE ELECTRIC 11-5
    HOOD 11-5
    KITCHEN 11-4
RECEIVING BOX 10-6
RECEPTOR SHOWER...................10-2, 15-6
RECESSED LIGHT FIXTURE 09-6
RECTANGULAR WALL TIE 04-2
RED BRICK 04-4
REDWOOD FENCE 02-13
    SIDING 07-5
REFRIGERATED CASE 11-4
REFRIGERATOR BLOOD 11-6

INDEX

REGLET 03-6
REINFORCING MASONRY 04-2, 04-8
REINFORCING STEEL REPAIRS 03-13
REMOVE FENCING 02-5
TOPSOIL 02-15
REPLACEMENT GLASS BLOCK 04-10
RESEED 02-15
RESIDENTIAL LOUVER 05-7
RETARDANT FIRE 06-12
REVOLVING DOOR 08-8
RIBBED BLOCK 04-7
RIDGE BOARD 06-4
ROOF 07-6
VENTILATOR 13-2
RIGID STEEL CONDUIT 16-2
URETHANE INSULATION 07-3
RISER BAND 11-2
MARBLE 04-11
WOOD 06-12
ROD GROUND 16-5
TRAVERSE 12-2
ROLL ROOF 07-3
ROLL-UP DOOR 08-6, 08-7, 13-2
ROOF ASPHALT 07-6
BUILT-UP 07-6
DRAIN 15-4
HIP 06-4
PANEL 13-3
SHINGLE 07-3
ROOFING REMOVAL 02-3
ROOFTOP AC 15-8
ROUGH-IN PLUMBING 15-5
ROUGH-SAWN SIDING 07-5
ROUND CAST BOX 16-5
QUARTER 06-10
RUBBISH REMOVAL 02-4
RUBBLE STONE 04-10

-S-

SAFETY GLASS 08-12
NET 01-4
SALAMANDER 01-5
SALES TAX 01-2
SALT PRESERVATIVE 06-11
SAND BED PAVER 04-6
BORROW 02-6
FINISH STUCCO 09-3
SANDING 09-9
SASH BEAD 06-10
SATIN FINISH LETTER 10-4
SCAFFOLDING 01-4
SCHEDULE-40 PIPE 02-11
SCHEDULING SOFTWARE 01-3
SCORED CLAY TILE 04-9
SCRATCH COAT 09-3
SCREED 03-8
SCREEN MOLDING 06-10
SCRUB SURGICAL 11-7
SECTIONAL DOOR 08-7
SECURITY SASH 08-8
SEDIMENT FENCE 02-9
SEEDING 02-15
SEH BOX 16-5
SELECT COMMON BRICK 04-4
SELF-DRILLING ANCHOR 05-4
SERVICE DOOR 08-6
SINK 15-6
SET RETARDER 03-8
S-FORM CONCRETE 03-6
SHAKE ROOF 07-4
SHEATHING ROOF 06-6
WALL 06-6
SHEET FLOOR 09-7
GLASS 08-12
METAL ROOF 07-6
PLASTIC 07-2
SHEETROCK FINISHING 09-4
SHELF 06-10
CHECKROOM 11-2
UNIT 10-6
SHINGLE CEDAR 07-4
METAL 07-3
REMOVAL 02-3
WOOD 07-4
SHOE MOLDING 06-10
SHOTCRETE 03-9
SHOWER COMPARTMENT 10-2
SHOWER DOOR 10-2
FAUCET 15-5
ROD 10-7
SHRUB MAINTENANCE 02-14
SIDE BEAM 03-2
SIDING 07-4, 07-5
SIGN INTERIOR 10-4
PLASTIC 10-5
SILL LIMESTONE 04-11
MARBLE 04-11
SLATE 04-11
WINDOWWALL 08-13
SIMULATED PEG FLOOR 09-6
SISTER JOIST 06-2, 06-3
RAFTER 06-5
SITE GRADING 02-5
SKYLIGHT 13-3
SLAB CHANNEL 03-12
CONCRETE 03-11
FORMWORK 03-3
MESH 03-7
REINFORCING 03-6, 03-7
SLATE 04-6
PANEL 04-11
SLIDE WATER 13-3
SLIDING FIRE DOOR 08-6
GLASS DOOR 08-7
WINDOW 08-10
SLIP COUPLING COPPER 15-2
SLOT ANCHOR 04-3
SNAPPED PAVER 04-6
SOAP DISH 09-5, 10-7
SOCIAL SECURITY 01-2
SOCKET METER 16-5
SODIUM LIGHT 16-7
SOFFIT REPAIR 03-13
SOFTENER WATER 11-5
SOFTWARE SCHEDULING 01-3
SOIL BORING 02-2
TESTING 01-3
SOLAR SCREEN BLOCK 04-8
SOLID ARMORED CABLE 16-4
BLOCK 04-6
CORE 08-3
WALL 04-4
SOUND DEADENING BOARD 09-3
SPALL PILE 03-13
SPANDREL GLASS 08-12
SPLIT GROUND FACE 04-7
RIB PROFILE 04-7
SPONGE RUBBER PADDING 09-8
SPREAD FOOTING 03-10
TOPSOIL 02-5, 02-14
SPREADING TOPSOIL 02-15
SQUARE COLUMN 03-3
TUB 15-4
STAGE BAND 11-2
THEATER 11-2
STAGING 01-4
STAINLESS COUNTER 12-2
DOOR 08-6
FLASHING 07-7
GRAB BAR 10-6
PARTITION 10-2
RAILING 05-7
SHELVING 10-6
STAIR LANDING 05-5
MARBLE 04-11
PAN 05-5
RAIL 05-6
TREAD CERAMIC 09-5
STALL SHOWER 10-2
STANDING-SEAM ROOF 07-6
STEAM GENERATOR MEDICAL 11-7
KETTLE 11-4
STERILIZER 11-7
STEAMER ELECTRIC 11-4
STEEL ANGLE 04-9
EDGING 02-15
GUTTER 07-7
LINTEL 05-4
PIPE 15-3
REPAIR 03-13
SHINGLE 07-3
SIDING PANEL 07-5
STEEL STAIR 05-5
STRUCTURAL 05-4
TANK 02-12
TOILET PARTITION 10-2
TREAD 05-6
WELDING 05-2
STEP LIMESTONE 04-11
MANHOLE 02-11
STEPPING STONE 02-15
STERILIZER GAS 11-7
MEDICAL 11-7
STILL WATER 11-7
STONE PAVER 04-6
STOOL MARBLE 04-11
MOLDING 06-10
SLATE 04-11
STOP MOLDING 06-10
STORAGE SHELVING 10-6
STORM DOOR 08-8
WINDOW 08-9
STRAIGHT CURB FORM 03-3
STRAND BEAM 06-8
STRAW BALE 02-9
STREET ELL COPPER 15-2, 15-3
STRIP CANT 06-5
CHAMFER 03-5
FLOORING 09-6
FLOURESCENT 16-6
SHINGLE 07-3
STRUCTURAL EXCAVATION 02-7
PLYWOOD 06-7
TUBE 05-4
STUCCO 09-3
LATH 09-3
STUD CLIP 09-3
METAL 05-5, 09-2
STUD REMOVAL 02-2
STUMP 02-5
STYROFOAM INSERT 04-7
SUB-BASE 02-8
SUB-FLOORING 06-6
SUBMERSIBLE PUMP 01-5
SUPERINTENDENT 01-2
SURFACE MOUNTED CLOSE 08-11
SURGICAL CHRONOMETER 11-8
LIGHT 11-7
SCRUB 11-7
SUSPENDED CEILING 07-2
CEILING REM 02-3
SUSPENSION SYSTEM 09-6
SWITCH DIMMER 16-6
SY-PINE DECKING 06-7
SIDING 07-5

-T-

T&G SIDING 07-5
TABLE LABORATORY 11-6
TACK BOARD 10-3
TANDEM BREAKER 16-5
TANK-TYPE EXTINGUISHER 10-5
TAPERED FRICTION PILE 02-9
TARPAULIN 01-4
TAX 01-2
T-BAR SYSTEM 09-6
TERMITE CONTROL 02-9, 10-3
TERRA COTTA 04-10
TERRAZZO RECEPTOR 10-2
REMOVAL 02-3
TEXTURE 1-11 07-5
THHN-THWN CONDUCTOR 16-4
THRESHOLD DOOR 08-11
MARBLE 04-11
THROUGH-WALL FLASHING 04-4
THWN CONDUCTOR 16-4
TIE MESH 04-4
TILE ACOUSTICAL 09-6
CLAY 04-9
QUARRY 09-5
REMOVAL 02-3
RESILIENT 09-7
STRUCTURAL 04-5
TIMBER FRAMING 06-3, 06-7
LANDSCAPING 02-16

TIMEKEEPER 01-2
TINTED GLASS 08-12
TISSUE DISPENSER 10-7
TOILET 15-7
    ACCESSORY 10-7
TOOTHBRUSH HOLDER 10-7
TOP PLATE 06-6
TOPDRESS SOIL 02-15
TOWEL BAR 10-7
    BAR CERAMIC 09-5
    DISPENSER 10-7
TRACK MOUNTED LOADER 02-6
TRAILER 01-3
TRAVERSE ROD 12-2
TRAVERTINE FLOOR 09-7
    FLOORING 04-11
TREAD ABRASIVE 05-6
    MARBLE 04-11
    METAL 05-5
    STEEL 05-5
    WOOD 06-12
TREATED LUMBER 02-16
    POST 06-7
TRENCHER 02-7
TRIM 09-9, 09-11, 09-12
    CARPENTRY 06-9
    CASING 06-9, 06-10
TRIPLE OVEN 11-3
TROFFER 16-6
TROWEL FINISH 09-3
    FINISH STUCCO 09-3
TRUCK 01-5
    CRANE 01-6
TRUSS REINFORCING 04-2
    WOOD 06-8
T-SERIES TILE 04-5
TUB BATH 15-4
TUBE STRUCTURAL 05-4
TUMBLESTONE AGGREGATE 04-6
TWIST LOCK RECEPTACLE 16-5

-U-

U LAMP 16-6
UNCASED BORING 02-2
UNDER SLAB SPRAYING 10-3
UNDERGROUND TANK 02-12
UNDERLAYMENT 06-6
UNDERSLAB DRAINAGE 02-13
UNEMPLOYMENT TAX 01-2
UNGLAZED FLOOR 09-7
    FLOOR TILE 09-4
UNICAM 16-7
UNIT HEATER 16-8
    KITCHEN 11-3
    MASONRY 04-8
UNRATED DOOR 08-2
URETHANE INSULATION 07-3
URINAL SCREEN 10-2
URN COFFEE 11-3
UTENSIL WASHER MEDICAL 11-7

-V-

VACUUM CARPET 09-8
    SYSTEM 11-2
VALLEY FLASHING 07-7
VALVE VACUUM 11-2
VANITY BATH 12-2
VAPOR BARRIER 07-2
VAULT DOOR 08-6
V-BEAM ROOF 07-6
VEGETATION CONTROL 02-9
VENEER ASHLAR 04-11
    BRICK 04-4
    GRANITE 04-11
VENETIAN BLIND 12-2
VENT CAP 02-12
    FOUNDATION 04-2
VERMICULITE 03-12
VERMONT SLATE 07-4
VERTICAL GRAIN FLOOR 09-6
    REINFORCING 03-8, 04-2, 04-8
VINYL FENCE 02-14
    SHEET 09-7
    SHEET FLOOR 09-7
    SHEETROCK 09-4
    SIDING 07-5
    TILE 09-7
    WALL COVERING 09-13
VISION GLASS 08-2
VISUAL AID BOARD 10-3

-W-

WAFERBOARD SHEATHING 06-7
WALK 03-11
WALL BASE 09-8
    BRICK 04-4
    CABINET 06-11, 12-2
    CAVITY 03-11, 04-2
    CLEANOUT 15-3
    FINISH 03-8
    FOOTING 03-4
    FORMWORK 03-5
    FURRING 05-4, 06-3
    HEATER 11-5
    HUNG SINK 15-6
    HYDRANT 15-6
    INSULATION 07-2
    LATH 09-3
    MASONRY REINFORCING 03-8
    MOUNTED FLAGPOLE 10-3
    MOUNTED HEATER 16-8
    PANEL 08-13, 13-3
    PLASTER 09-3
    RAILING 05-6
    REMOVAL 02-4
    SHEETROCK 09-4
    STONE 04-10
    TIE 04-2
    TILE 09-4, 09-5

WALLS 09-11, 09-12
    AND FLAT SURFACE 09-9
WALNUT DOOR 08-5
WARDROBE 12-2
    LOCKER 10-5
WASHER EXTRACTOR 11-2
WASHING MACHINE BOX 15-7
WASHROOM FAUCET 15-5
WASTE RECEPTACLE 10-7
WATCHMAN 01-2
WATER BATH 11-6
    BATH LABORATORY 11-6
    REDUCING ADMIXTURE 03-8
    SLIDE 13-3
    SOFTENER 11-5
    STILL 11-7
WATER-RESISTANT SHEET 09-4
WAX FLOOR 09-6
WEARING SURFACE 02-10
WELDED RAILING 05-6
WELDING 05-2
WET BED TILE 09-4
    RUB 03-9
WHEEL BARROW 02-7
    CHAIR PARTITION 10-2
WHEELED EXTINGUISHER 10-5
WHITE CEDAR SHINGLE 07-4
WIDE STILE DOOR 08-13
WINDOW 08-8, 08-9
    FRAME 08-10
    REMOVAL 02-3
    SILL 04-5
    SILL MARBLE 04-11
    STEEL 08-8
    STILL CERAMIC 09-5
    STOOL MARBLE 04-11
WINDOWS 09-11
WINDOWWALL 08-13, 08-14
WIRE MESH 03-7, 03-11
WOOD CANT 06-5, 07-6
    CHIP MULCH 02-15
    COLUMN 06-12
    DECKING 06-7
    DOOR REMOVAL 02-2
    FENCE REMOVAL 02-5
    FLOOR REMOVAL 02-3
    FRAME 08-5
    HANDRAIL 05-7
    PILE 02-10
    PLATE 06-6
    RISER 06-12
    SHELVING 06-10
    SOFFIT 06-5
    STUD REMOVAL 02-2
    WINDOW REMOVAL 02-4
WOOL CARPET 09-8

-X-

X-RAY EQUIPMENT 11-7

-Z-

Z-TIE WALL 04-3